Saving Lives, Step by Step: Robust Techniques for Landmine Detection

Perry

Contents

Chapter 1

Introduction

Today, landmines and other unexploded ordnance (UXO) are one of the greatest curses of modern times, the legacy of war killing and maiming innocent people everyday. Ever since Landmine Monitor 2020 began global tracking in 1999, at least 130 000 civilian casualties (of which 48% are children) have been reported, caused by such landmines [Monm]. It is inevitable that many more occurrences went unrecorded due to the lack of national monitoring systems in some affected countries. Landmines not only pose a threat to the lives and well-being of the population, but also hinder post-conflict reconstruction efforts and the economic development of a nation. Detection and clearance of landmines is a very risky process, which requires high standards on technology and methodology.

Ground penetrating radar (GPR) is one of the leading sensing modalities for landmine detection [Dan04; Zou+02]. It is a nondestructive technology that transmits electromagnetic radiation into the ground and measures the reflections arising due to the dielectric contrast of targets embedded in the soil, thereby permitting detection and localization of both metallic and non-metallic landmines. A forward-looking ground penetrating radar (FL-GPR) offers the potential to detecting landmines with reduced risk to the operator [SL03; Wan+07; JLZ12; Com+17; Com+21]. This is accomplished by a large standoff distance between the detector and the targets, rendering it an attractive option for humanitarian demining processes compared to a conventional downward-looking radar that incurs the chance of disturbing and damaging the target scene. However, a challenge of using FL-GPR is that the illuminating signals and the reflected signals experience substantial attenuation owing to the near cancellation of the direct and ground-reflected waves. Furthermore, the interface roughness and subsurface clutter, which are usually highly non-stationary, can be detrimental to the target detection performance of an FL-GPR system. In order to provide reliable detection, robust methods that are insensitive to changes in surface characteristics are highly desirable.

This book aims to design robust signal processing techniques for landmine detection with FL-GPR. The robust techniques have to be able to provide a guaranteed performance in the case of anomalies in the mine contaminated environments.

1.1 Motivation

For many decades, as a cheap and easy-to-use alternative, landmines have become a favourite weapon in civil wars and insurgency wars, used by both the security forces and guerillas. The use of landmines in battlefields violates international humanitarian law, which intends to protect people who are no longer or are not participating in armed conflicts. Many years after a conflict has ended, landmines continue to pose threats to civilians due to little or no effort being made to clear them from the affected area. They limit freedom of movement and deny access to basic human needs, which hinders post-conflict reconstruction efforts and the implementation of the sustainable development goals (SDGs) of a country.

Demining and clearance of landmines is a dangerous activity that is obliged to follow high standards in its methodoloy and technology [Ima]. In humanitarian demining, the clearance activity is conducted not for the military, but for the benefit of civilians. It aims to reduce risks for the deminers and civilians as much as possible. The current use of manual detection tools contributes to the relatively very slow progress of humanitarian demining [Monm; Mong]. As an alternative technology, an FL-GPR system has garnered increasing research interest in the last few decades due to its ability to provide stand-off detection of buried targets [SL05; TWS10; Cat+15; Com+20].

In the application of FL-GPR, detection methods that are insensitive to the changes of surface characteristics of the investigated environments are highly desirable in order to provide reliable detection [Com+17; Kos+02; SL03]. Due to the property of having a guaranteed performance under various conditions, robust signal processing techniques can potentially contribute to increasing the performance and usage of FL-GPR for detecting landmines, saving lives and modernizing efforts in the battle against these life-restricting devices once and for all.

1.2 State-of-the-Art

FL-GPR systems have primarily been developed for military operations [RRE05; Wan+05; Res+07; Pag+15]. An FL-GPR system that works in the range of 400 MHz–4 GHz using a stepped-frequency continuous-wave (SFCW) signal was developed by Planning Systems, Inc. (PSI) [Wan+05]. An FL-GPR prototype was developed by the Stanford Research Institute (SRI) to operate with a stepped frequency signal covering 500 MHz–3 GHz [RRE05]. Synchronous impulse reconstruction (SIRE) radar is an ultra-wideband forward-looking radar operating at 300 MHz–3 GHz bandwidth range designed by the US Army Research Laboratory (ARL) to detect landmines and improvised explosive devices (IEDs) [Res+07]. The Lawrence Livermore National Laboratory (LLNL) has developed a multistatic GPR system mounted on a moving vehicle capable of real-time imaging and object detection [Pag+15].

Considerable work has been carried out to overcome the challenges of FL-GPR [SL03; SL05; Wan+07; TWS10; JLZ12; LDS14; Com+17; Com+18]. An ambiguity function based detector was proposed in [SL03], which exploits the time–frequency characteristics of target and clutter scattering. This motivated the use of a wavelet packet transform and a neural network classifier in [SL05] to increase the detection performance. In [Wan+07], frequency subband processing and polarimetric features, among others, were used to improve target detection performance. In [TWS10], an adaptive multi-transceiver imaging technique is proposed that effectively integrates imaging information obtained through platform motion. Feature extraction of the bistatic scattering images from a multiple-input-multiple-output radar system is proposed in [JLZ12] to differentiate targets from clutter. A time-reversal imaging algorithm is proposed in [LDS14], which employs three-dimensional finite-difference time-domain modeling to characterize the scattering from the ground surface and buried targets. The method proposed in [Com+17] forms a multi-view tomographic image as a reference for performing a statistical parametric test to detect shallow-buried landmines in the presence of a rough ground surface. Furthermore, an adaptive approach is proposed in [Com+18] to reduce the false alarm rate by iteratively updating the estimated parameters of the clutter and target distributions under changing viewpoints.

While all the above works considered the influence of clutter in the detection performance, to the best of our knowledge, none of them focused solely on the distribution uncertainty of clutter and target signatures in the beamformed FL-GPR image. Considering the applications of landmine detection using FL-GPR, detection techniques that can be robust and provide a guaranteed performance under such uncertainty are highly desirable.

1.3 Contributions

The contributions of this book are as follows:

- **Uncertainty Distribution in Radar Image:** Distributions of targets and clutter in the FL-GPR imagery are examined. Confidence bands around density estimates are constructed via bootstrap resampling and kernel density estimation methods. These bands can be used to observe the distributional uncertainty of targets and clutter of the radar image.

- **Robust Technique for Landmine Detection:** Robust techniques that offer a guaranteed performance in the case of deviations in the distributions are developed for landmine detection. The detectors are designed such that they minimize the maximum error probability over all feasible distributions. They improve the performance of classical detectors based on nominal models.

- **Dependence Structure via Copula Model:** The design of the landmine detectors is further extended to capture the statistical dependence between FL-GPR images via a copula-based model. The usage of different copula models is investigated and analyzed. Incorporating dependence structure with a well specified copula offers optimized performance for both classical and robust detectors.

1.4   book Overview

The doctoral book outline is as follows. Chapter 2 describes the global challenge of extensive landmine contamination, current detection tools and alternative sensing modalities for landmine detection. It further introduces the configuration and imaging approach of FL-GPR.

Chapter 3 considers the problem of landmine detection in FL-GPR imagery. Simple threshold and likelihood-ratio test (LRT) detectors are examined and applied to multiple radar images obtained from various viewpoints of the region of interest (ROI). A parametric family of distributions is employed to obtain a nominal model of targets and clutter distributions in the radar image.

Chapter 4 introduces uncertainty distribution models for the FL-GPR image. Distributional uncertainties of targets and clutter in the radar image is investigated. Density bands are constructed to capture feasible distributions of the radar image. Detection techniques are designed to be robust against the deviations of distributions.

In Chapter 5, the detection techniques are further extended to capture the statistical dependence between images from two consecutive viewpoints via a copula-based model. Different copula density functions are investigated in terms of their effectiveness in incorporating dependence structures between multi-view images.

Finally, conclusions and outlook for future work are given in Chapter 6.

Chapter 2

Landmine Problem and Existing Demining Approaches

In this chapter, the global challenge of extensive landmine contamination is introduced. A description of existing detection modalities for humanitarian demining as a critical part of mine action[1] operations is provided to put forward the view of further research described in the book.

2.1 Global Landmine Contamination

Since the end of World War II, landmines have been used in many conflicts around the globe. As a cheap and easy-to-use alternative, landmines have become a favourite weapon in civil wars and insurgency wars, used by the security forces and guerillas and violating international humanitarian law, which intends to protect people who are not or are no longer participating in armed conflicts. Landmines contain explosive substances which are designed to disable or even destroy enemy targets. They are deployed under or on the ground surface, and can be activated by either

[1]The term 'mine action' refers to groups of activities which aim to reduce the social, economic and environmental impact of landmines and explosive remnants of war (ERW) [Unic]

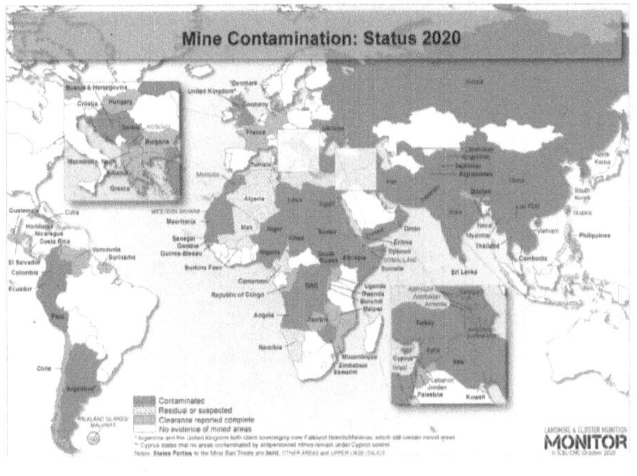

FIGURE 2.1: Contaminated countries (status as of October 2020).
Source: International Campaign to Ban Landmines [Inta]

a person walking or a vehicle driving over them, triggering the firing mechanism. There are two categories of landmines: anti-personnel (AP) mines and anti-tank (AT) mines. AP mines are a type of landmine designed to detonate from the proximity or contact of a person. An AT mine is typically larger in size and has much heavier explosive materials than AP mines. AT mines are made to disable or destroy civilian and military vehicles and their occupants. Their explosions are triggered by the proximity of or contact with a vehicle as opposed to a person [Swi]. After the ends of conflicts, landmines pose a serious problem due to little or no effort being made to clear them from the area. They continue to pose threats to civilians many years after a conflict has ended, restricting people in carrying out their daily activities. They limit freedom of movement and deny access to food, water and other human basic needs, which hinder post-conflict reconstruction efforts and the implementation of the SDGs of a country [Unia].

Fig. 2.1 shows contaminated countries around the world from the most recent report published by the Landmine Monitor [Monm]. As of October 2020, there are sixty states and other areas contaminated by AP mines. Massive contamination, which is defined as more than $100\,\mathrm{km}^2$ of land area by the Landmine Monitor, exists in ten countries: Afghanistan, Yemen, Ukraine, Bosnia and Herzegovina (BiH), Turkey, Cambodia, Thailand, Croatia, Iraq and Ethiopia. Mauritania, which declared itself mine-free in 2018, announced a new contamination area in its suspected territory in 2019 a legacy of the Western Sahara conflict from the 1970s. Three countries–Algeria, Kuwait and Nicaragua–are required to clarify the extent of residual contamination in their area. Five countries–Cameroon, Mali, Burkina Faso, Nigeria and Tunisia–have to provide information about suspected or new contaminated areas of improvised

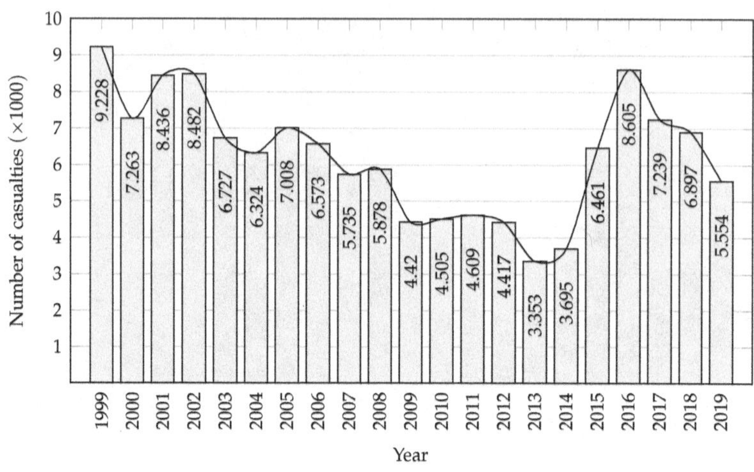

FIGURE 2.2: Global casualties from 1999 to 2019, derived from [Monj; Monk; Monl; Monm]

mines. The current status of contamination does not differ very much compared with the status from two decades ago, which were sixty-five countries confirmed to be mine-contaminated as of October 2004 [Mona], even though mine-clearance activities have been conducted through international donors for mine action every year. Ongoing use and production of mines in many conflicts contribute to this global problem. For example, from 2004 to 2009, insurgents in Colombia, India, Myanmar and Pakistan used AP mines every year [Monb]. The security forces of Myanmar continued to use AP mines during the fight with ethnic armed groups in 2017 [Wen]. Colombia, as one of the most contaminated countries in the world, has thousands of landmines across many areas in the country which were planted by the left-wing guerilla group during its five-decade fight with the government [Mol]. Since 2014, eleven countries–Vietnam, China, Cuba, South Korea, India, Singapore, Iran, Myanmar, Russia, North Korea and Pakistan–have been listed as potential producers of AP mines by the Monitor because they have not refused future production. From 2018, insurgents have continued producing mines in Afghanistan, Colombia, Myanmar, Pakistan and Yemen.

There is no precise figure for the number of total landmines in the ground; however some estimates suggest 110 million mines scattered across the globe with more being deployed every year [BBC; Hum]. The Landmine Monitor 2018 revealed an estimate of 50 million AP mines stockpiled around the world [Bro; Monk]. The total of 131 401 casualties caused by mines and other UXO has been recorded by the Monitor since its global tracking began in 1999. Fig. 2.2 shows the number of recorded casualties per year over two decades from 1999 to 2019. A high number of casualties was recorded in 2016, of which at least 8605 people were killed or injured. The majority of casualties (78%) were civilians with children making up 42% of total civilian casualties. This record has marked the highest casualty report since 1999, which reflects an average of $\approx$ 23 people killed or injured per day in 2016. The lowest level of casualties was recorded in 2013 with a global total of at least 3353, which equals to an incidence rate of $\approx$ 9 casualties per day. A sharp increase in the number of casualties occurred in 2015, where clearance funding hit its ten-year low [Moni]. Armed conflicts in Ukraine, Yemen, Libya and Syria caused more severe conditions for the civilian victims and contributed to the sharp increase in casualties of 2015, which was 74.8% higher than the previous year. While remaining high, the number of recorded casualties has consistently decreased from 2016 to 2019. However, it is certain that many more casualties went unrecorded due to lack of national monitoring systems in severely affected countries [Monm].

2.2 Humanitarian Demining Action

Demining is the operation of removing or clearing landmines from a contaminated area. In humanitarian demining, the clearance activity is conducted not for the military, but for the benefit of civilians. It aims to reduce risks for the deminers (operators) and civilians as much as possible. Humanitarian demining is a main pillar in mine action, groups of complementary activities which aim to reduce human suffering and the negative socio-economic and environmental impacts of mines to people living in affected countries and regions [Unic; Ger]. As stated by the United Nations mine action service (UNMAS), mine action consists of five complementary pillars–clearance, mine risk education, victim assistance, advocacy and stockpile destruction–which together support the vision of a world free from the threat of landmines and UXO.

One of the major global non-governmental organizations (NGOs) in humanitarian demining is the international campaign to ban landmines (ICBL), whose objective is to achieve a mine-free world. The ICBL succeeded in proposing the Mine Ban Treaty, known as the Ottawa Treaty, an international agreement that bans AP mines around the globe. Since declared open for signing in December 1997, the treaty has been signed by 164 states [Monm]. The member states of the mine ban treaty commit to the following agendas: (1) never using AP mines or producing, stockpiling or transferring them, (2) clearing mine-contaminated areas under their jurisdiction within ten years, (3) destroying all mines in their stockpiles within four years, (4) conducting mine risk education and ensuring civilians are excluded from contaminated areas and (5) offering assistance to other state members which lack resources for mine clearance. Thirty-two countries have not signed the treaty, including China, India, the USA, Saudi Arabia and Russia. The ICBL continue together with the United Nations (UN) to lobby the non-signatory countries to adopt the treaty [Intb; Unib].

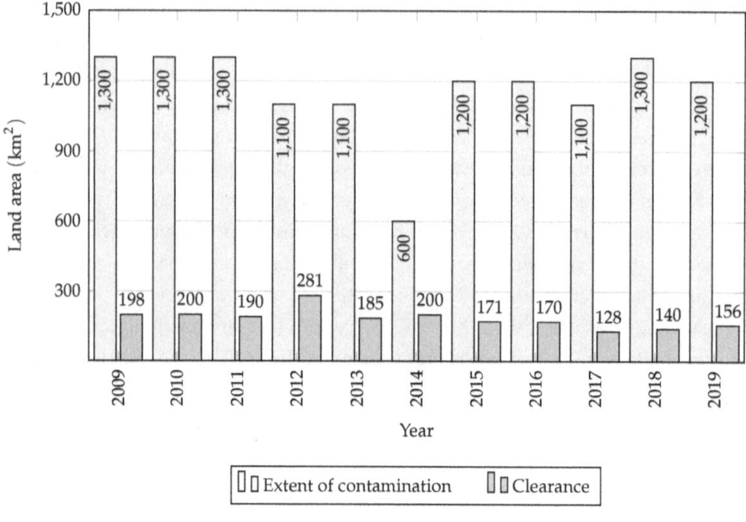

FIGURE 2.3: Completed clearance of landmine areas and extent of contaminated areas from 2009 to 2019, derived from [Monb; Monc; Mond; Mone; Monf; Mong; Monh; Moni; Monj; Monk; Monl; Monm]

Besides UNMAS, there are other international organizations involved in mine action, i.e., the Geneva International Centre for Humanitarian Demining (GICHD) and the national Mine Action Centres (MAC) in Sri Lanka, Cambodia, Croatia, Laos, etc. Other demining NGOs such as The Halo Trust, Norwegian People's Aid, DEMIRA (Deutsche Minenräumer e.V.) and Mines Advisory Group are also involved in the removal of landmines and UXO for humanitarian reasons. On behalf of the UNMAS, the GICHD manages and coordinates the process of updating the international mine action standards (IMAS), which include a set of standards for humanitarian demining procedures. Both governmental and non-governmental organizations actively review these standards to take into account new developments in demining methodology and technology. The IMAS aims to provide a common and consistent procedure for mine clearance activity in a safe and efficient manner.

Fig. 2.3 shows the rate of landmine clearance and the extent of contamination per year recorded by the Landmine Monitor only from 2009 to 2019. The highest clearance rate was in 2012 with approximately 281 km^2 land area cleared of landmines, while the lowest clearance rate was at 128 km^2 in 2017. Over the past eleven years, approximately 2019 km^2 of mine-contaminated land has been cleared. This clearance operation has had little impact given the extent of contamination every year shown in Fig. 2.3. The extent of contamination is derived only from countries which have contamination of approximately 100 km^2 per year. Reported countries having extent of contamination lower and higher than 100 km^2 were not included in Fig. 2.3.

The extent of contamination varies between three to nine times more than the clearance rate, which makes a mine-free world an unachievable goal. The rate of mine clearance is affected mainly by two factors: the availability of international financial support for mine action and the use of current technology for demining. If demining operations remain the same as the current status and no more mines laid, an estimate of a thousand years is expected for clearing all existing mines [Lom19].

2.3   Sensing and Detection Modalities

Detection and clearance of landmines is a high risk operation which can put the operator's life at stake. In humanitarian demining, the safety of the operator is a high priority. This section presents current detection tools and alternative sensing modalities for landmine detection.

2.3.1   Metal Detector

A widely used and common tool for landmine detection in humanitarian demining is a metal detector. Deminers continue to use metal detectors as the primary tool for detecting landmines. The high reliance on metal detector use in humanitarian demining is due to several factors: good understanding and easy operation of the detectors, relatively inexpensive technology and current procedures aimed solely at metal-free ground, although mines with zero metal content exist. However, metal detectors have significant false alarm rate (FAR) due to the presence of metal clutter such as shrapnel or fragments of bomb and shell cases in the minefield, which ends up wasting a lot of time and effort for excavating non-lethal items. Deminers ideally should ignore those items and focus more on explosive devices in the area of interest.

In the demining process, five stages are defined by the UN [Ima]: (1) finding mine-contaminated area (2) determining the locations of mines in the area of interest (3) confirming the individual mines (4) removing all suspect objects (5) confirming that the area is safe and ready to open for civilians. Metal detectors are used in the first, second, third and fifth of the defined stages. For humanitarian demining, the process is conducted to detect, clear and destroy all landmines in a non-military region. Before investigating the ROI, a basic operational test of metal detector has to be performed to check its general condition and functionality and adjust it to the interrogated ground condition. The test is repeated for any changes in underlying conditions during the operating session. It is also necessary to check the manual of the metal detector provided by the manufacturer before setting it up. These comprehensive checks ensure that deminers operate the detector correctly for the safety of their work. An example of a minimum series of prior checks before operation is as follows [Gue+03]: (1) inspect the detector's condition in general such as battery, all wire connections, any visible damage, etc., (2) test the functionality of the detector, which takes for a few minutes, (3) adjust the detector's parameters to the local ground conditions, (4) check whether the detector can detect known targets in similar ground conditions, (5) repeat the detector's functionality test if the condition of the mine area have changed, (6) dismantle the metal detector and check again its general condition.

During the survey, every signal generated by the metal detector has to be considered by deminers as a response to a mine even though its strength varies and can not be reliably distinguished from background signals. The deminer can be more confident to suspect a mine existed underground if a signal obtained at one spot exists in parallel with other signal responses from targets already positively identified. In every inspection, any changes in the metal detector response should be considered as a potential mine area. In order to locate mines, a deminer needs to take considerable effort using the metal detector. In addition, many minefields are within areas of previous conflicts which contain significant metal contamination. This condition leads to a significant false alarm (FA) rate , which causes deminers to often find non-lethal items–up to a hundred for every positively identified target [JL03]. This contributes to the long operation time and high clearance cost and affects the deminer's safety.

2.3.2 Ground Penetrating Radar

As an alternative sensing technology, GPR can be used for landmine detection. GPR has been used in many applications in a number of fields, such as in geology for soil study and in archeology for detecting and mapping objects of archeological interests.

The capability to detect plastic landmines via radar is one of several reasons why organizations around the world are persuing research in the field of GPR [Lew+06; Dan04; Tes13; Lom19; Hu18]. Combining a GPR with traditional metal detectors, the FAR can be reduced considerably and the detection rate increased [Kop07]. GPR systems work in a similar way to radar. Both technologies are based on the principle of an electromagnetic wave that is first emitted, then reflected by one or more objects and finally captured and measured. One advantage of GPR, however, is that no ground contact is required to measure the wave reflection. One or more antennas pick up the wave reflections and make it possible to determine the differences in dielectric constants and other electromagnetic properties that occur at the material boundaries of underground objects and the air–earth boundary. In the context of landmine detection, of course, the material boundaries between the target objects and the soil surrounding them are of particular interest.

When analyzing the wave reflection, the GPR system can determine the shape and position of the underground objects, provided that a suitable analysis method has been selected [JL03; Zou+02]. Finding the target objects is made more difficult if the soil has a similar electrical conductivity to the target objects, i.e. the material boundaries are less pronounced. A metal mine in a poorly conductive soil creates an easily recognizable boundary, while a plastic mine in a sandy soil is difficult to identify [Mac+03]. Heterogeneous subsurface features such as roots, stones and water sacs can generate disruptive signals [JL03]. Other aggravating factors can be: rain, moisture or conductivity of the soil and heterogeneity in the nature of the landmine itself.

The ground surface itself reflects a large part of the signal and usually also has the largest dielectric constant. This effect is known as "ground bounce" [Mac+03] and is particularly disruptive when plastic mines are to be discovered at shallow depths. However, it can be significantly reduced by GPR systems with a higher bandwidth and very short pulses with negligible ringing, which leads to an increased resolution [JL03]. The limitations and challenges described here are the primary cause of the FAR for GPR.

The choice of frequencies used also plays a role. Electromagnetic waves of low frequencies (in the range of several hundred MHz) can penetrate deeper into the ground, but only provide a comparatively low image resolution. Waves of high frequencies (in the low GHz range), on the other hand, provide higher image quality, but do not penetrate very deeply into the ground. As a compromise, the operating frequencies are usually somewhere in between. GPR systems are usually only operated with weak power, which makes them harmless for the user [Bru+99].

2.3.3   Forward-Looking Radar

The aforementioned GPR system operates in downward-looking (DL) mode. The DL-GPR can only detect buried targets reliably when the antenna is close or directly above the ground surface in a variable range of 0.15–0.5 m [Ben+17]. In the last few decades, there has been an increasing interest in developing techniques to detect buried targets from a stand-off distance. This has led to a new class of GPR which is known as forward-looking GPR (FL-GPR). The FL-GPR offers the potential of detecting landmines and UXOs with reduced risk to the operator, owing to the large distance between the detector and the targets. Fig. 2.4 illustrates an FL-GPR vehicle-based system with an antenna (sensor) array mounted on top. The radar system interrogates area in front of the vehicle from a stand-off distance.

FL-GPR systems have primarily been developed for military operations [Kos+02; JLZ12; Sol+17; CCM18; LDS14; RRE05; Cat+21]. However, these systems should also be potentially considered for the humanitarian demining operation where the safety of the operators is a high priority. In sensing using FL-GPR, the area of interest is always in front of the radar platform at a considerable large and safe distance for the safety of the operators. An FL-GPR system which works in the range of 400 MHz–4 GHz SFCW signal was developed by using a PSI [Wan+05]. An FL-GPR prototype was developed by the SRI to operate with a stepped frequency signal covering 500 MHz–3 GHz [RRE05]. SIRE radar is an ultra-wideband forward-looking

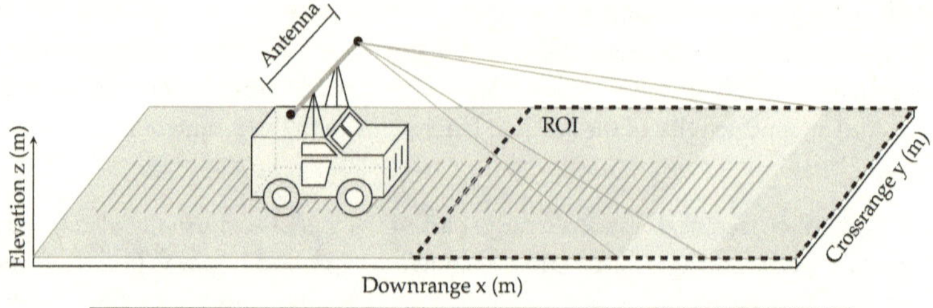

FIGURE 2.4: FL-GPR vehicle-based system with antenna array mounted on top. Parallel lines on the ground indicate sensor array positions.

radar designed by the ARL to detect landmines and IEDs [Res+07]. The radar system uses two transmitters and sixteen receivers and operates at 300 MHz to 3 GHz bandwidth range. The configuration of the SIRE radar system is considered in this book.

Due to the adopted data-acquisition geometry, there may be reduced wave-coupling in the ground in FL-GPR imaging. A rough surface and subsurface clutter also have a negative effect on the detection of the target objects. Therefore, computational electromagnetic tools based on advanced scatter models have been developed to predict radar performance under various environmental clutter conditions [LDS14]. Real-time three-dimensional (3-D) modeling of the moving-platform FL-GPR scattering in rough terrain was reported in [Taj+19]. Microwave tomographic approaches for FL-GPR imaging were developed in [Sol+17]. In particular, in [Sol+17] the imag-ing performance of the beamforming and microwave tomographic approaches was analyzed and compared under flat and rough terrain conditions. Likelihood-ratio test-based detectors were considered in [Com+21] and [Pam+20b] to detect landmines and targets with low signature, while a coherence factor-based suppression of rough surface echoes was recently reported in [Com+20].

2.4 Forward-Looking Radar Data

The FL-GPR data used in this work was simulated using the near-field finite-difference time-domain (NAFDTD) software package developed by the US Army Research Laboratory (ARL) [Dog12]. The radar data assumes a ground-based FL-GPR imaging system, illustrated in Fig. 2.4.

2.4.1 Radar Configuration

A vehicle-based FL-GPR system is considered, which is equipped with a 2 m wide antenna array composed of 2 transmit and 16 receive elements. The antenna array is mounted on top of a vehicle at the approximate elevation of $z = 2\,\text{m}$, as illustrated in

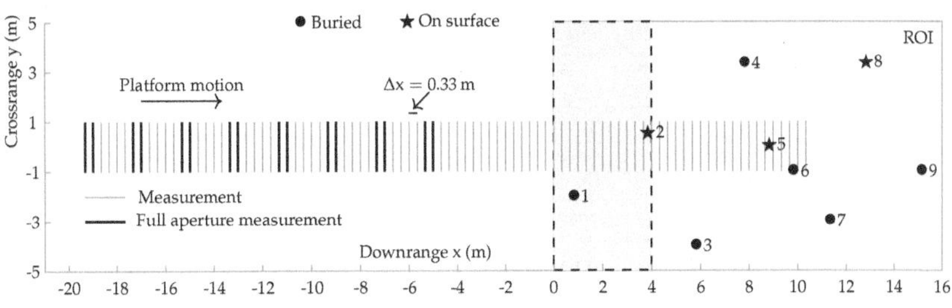

FIGURE 2.5: Top view of the FL-GPR measurement. Eight full aperture measurements are integrated to image the first segment (dashed line rectangle), corresponding to the farthest (first) viewpoint.

Fig. 2.4. The transmit elements are placed on the extreme ends of the array of receive elements, which are uniformly distributed along the y–direction. In forward-looking mode operation, the radar platform moves along the x–direction to sense the ROI at grazing angle $\theta_g \in [5°, 20°]$ approximately, with a stepped frequency signal covering the 0.3–1.5 GHz band in 6 MHz increments. In each array position of the moving radar platform, one of two transmit elements is operated and all receive elements simultaneously record the reflected signals from the ROI. By alternating left/right transmit element operation over two consecutive array positions, full aperture measurements are generated, which corresponds to 32 transmit–receive channels. This radar configuration mimics the US Army Research Laboratory's experimental SIRE radar testbed [Res+07].

Fig. 2.5 depicts the consecutive top-view measurement configuration considered in the simulation. The radar platform moves forward along the downrange from $x = -20\,\mathrm{m}$ to $x = 11\,\mathrm{m}$, providing different viewpoints of the 16 m × 10 m ROI. A complex interrogation scene is considered, which is a rough surface environment containing nine plastic and metallic targets. The types of targets are detailed in Table 2.1. Six targets are buried at a depth of 3 cm, five of which are metallic landmines {1, 3, 4, 6, 7} and one is made of plastic {9}. The remaining targets, two plastic landmines {2, 8} and a metallic one {5}, are placed on the surface. The considered on-surface and shallow-buried target deployments mimic realistic scenarios, typical of AP landmines and/or AT landmines laid by specialized minelayer vehicles [The15]. Plastic landmines are characterized with a relative dielectric constant $\varepsilon_r = 3.1$ and conductivity $\sigma = 2\,\mathrm{m\mho/m}$, which is representative of a minimum-metallic mine [M14]. The ground is modeled as a dielectric medium which is non-dispersive, non-magnetic, and homogeneous with $\varepsilon_r = 6$ and $\sigma = 10\,\mathrm{m\mho/m}$. These electrical properties are typical of medium-dry soil [Itu]. The surface roughness is described as a two-dimensional Gaussian random process parameterized by the root mean square height h_{rms} and the correlation length l_c [Com+17]. Both low ($h_{rms} = 0.8\,\mathrm{cm}$, $l_c = 14.26\,\mathrm{cm}$) and high ($h_{rms} = 1.6\,\mathrm{cm}$, $l_c = 14.93\,\mathrm{cm}$) surface roughness profiles are

TABLE 2.1: Complete list of targets.

No.	Target	No.	Target
1.	Buried metallic AP landmine	6.	Buried metallic 155–mm shell
2.	On-surface plastic AP landmine	7.	Buried metallic 155–mm shell
3.	Buried metallic 155–mm shell	8.	On-surface plastic AP landmine
4.	Buried metallic AT landmine	9.	Buried plastic AT landmine
5.	On-surface metallic AT landmine		

considered. A total of 90 array positions spaced $\Delta x = 0.333$ m apart are considered, whose projections on the x–y plane are depicted as parallel lines in Fig. 2.5. Two adjacent array positions realize a full aperture measurement. Between two consecutive activations of the same transmit element, a fixed displacement of 0.666 m occurs along the x–direction. For each array position, the electromagnetic scattering problem is solved numerically by the inverse scattering model [Per14; Sol+17; Dev12].

2.4.2 Imaging Approach

The processing of FL-GPR data is realized with a tomographic imaging approach, which allows image formation from scattered signals to be stated as in an inverse scattering problem. A simple sketch of an imaging scenario is illustrated in Fig. 2.4, which involves two media: medium 1 (the air) where the radar platform is located and medium 2 (the soil) where targets of interest reside. The targets of interest can be modeled as materials with their electrical properties denoted by $\varepsilon(r)$ and $\sigma(r)$, with $r = (x, y, z) \in \Omega$, which are the coordinates where they are located at the interrogation area or the imaging domain (ID) Ω inside medium 2. The electrical properties of the ID are parameterized by the dielectric constant $\varepsilon'(r)$ and the electrical conductivity $\sigma'(r)$.

Tomographic imaging aims at retrieving an unknown contrast function $\chi(r)$, which represents the variations of the electrical features of the targets with respect to the electrical features of the interrogated medium. The function is mathematically defined as [Amb+19]

$$\chi(r) = \frac{\varepsilon(r) - j\sigma(r)/\omega\varepsilon_0}{\varepsilon'(r) - j\sigma'(r)/\omega\varepsilon_0} - 1 \tag{2.1}$$

with $\omega = 2\pi f_r$ being the radar operating frequency with the frequency range $f_r = [f_{rmin}, f_{rmax}]$. The interrogation area is sensed by a transmit/receive antenna array located at a given height on domain Γ in medium 1. For each measurement at $r_e \in \Gamma$, the scattering phenomenon can be described by a pair of fundamental equations of the radiated and measured electrical fields [Per14; Pas10] respectively as

$$E_{tot}(\boldsymbol{r},\omega) = E_{in}(\boldsymbol{r},\omega) + k_p^2 \int_\Omega G_{in}(\boldsymbol{r}',\boldsymbol{r},\omega)\chi(\boldsymbol{r}')E_{tot}(\boldsymbol{r}',\omega)d\boldsymbol{r}' \tag{2.2}$$

$$E_{sc}(\boldsymbol{r}_e,\omega) = k_p^2 \int_\Omega G_{ex}(\boldsymbol{r}_{ex},\boldsymbol{r},\omega)\chi(\boldsymbol{r})E_{tot}(\boldsymbol{r},\omega)d\boldsymbol{r} \tag{2.3}$$

where k_p is the propagation constant and $\boldsymbol{r},\boldsymbol{r}' \in \Omega$. The total radiated field E_{tot} is defined by (2.2), which is the sum of the incident electric field E_{in} inside Ω when the targets are absent and the integral entity which defines the mutual interactions with an elementary source inside Ω. The scattered electric field E_{sc} is measured at $\boldsymbol{r}_e$ in medium 1. G_{in} and G_{ex} are the dyadic Green's functions, which can be interpreted as the impulsive electric field inside and outside the ID radiated by elementary sources located at $\boldsymbol{r}' \in \Omega$ and $\boldsymbol{r} \in \Omega$, respectively.

Note that equations (2.2) and (2.3) show the non-linearity relationship between the unknown contrast function χ and the scattered electric field E_{sc}. Although the scattered field is linearly related to the product between the total electric field and the contrast function as seen in (2.3), the total field itself depends on the contrast function by (2.2). Under the Born approximation [Cat+15; LS03], the total field inside the targets can be assumed to be almost equal to the incident field, i.e, $E_{tot} = E_{in}$, which makes (2.2) linear. By modeling the transmit antenna as a short dipole antenna with length l_d and charged with a magnitude I_0, the incident electric field can be written as

$$E_{in}(\boldsymbol{r},\omega) = -j\omega\mu_{sp}I_0 l_d G_{in}(\boldsymbol{r}',\boldsymbol{r},\omega) \tag{2.4}$$

where μ_{sp} is the permeability constant of free space. For the transmit antenna current moment $I_0 l_d$ assumed to be equal to 1 A–m, the scattered electric field model in (2.3) becomes

$$E_{sc}(\boldsymbol{r}_e,\omega) = -j\omega\mu_{sp}k_p^2 \int_\Omega G_{ex}(\boldsymbol{r}_e,\boldsymbol{r},\omega)G_{in}(\boldsymbol{r}',\boldsymbol{r},\omega)\chi(\boldsymbol{r})d\boldsymbol{r}. \tag{2.5}$$

Since the targets of interest are shallowly buried and the radar platform positioned at long stand off distance, the dyadic Green's functions can be approximated as the propagation model in a homogeneous medium with electrical features of free space [LDS14; Sol+17]. The dyadic Green's function for free space is of the form

$$G(\boldsymbol{r},\boldsymbol{r}_0,\omega) = \left[\boldsymbol{I} + \frac{1}{k_p^2}\nabla\nabla\right]g_{ar}(\boldsymbol{r},\boldsymbol{r}_0,\omega) \tag{2.6}$$

where $\boldsymbol{I}$ is an identity matrix, $\boldsymbol{r}_0$ is equal to $\boldsymbol{r}_e$ or $\boldsymbol{r}'$ and $g_{ar}(\cdot)$ is the scalar Green's function given by

$$g_{ar}(\boldsymbol{r},\boldsymbol{r}_0,\omega) = \frac{1}{4\pi}\frac{e^{jk_p|r-r_0|}}{|r-r_0|}. \tag{2.7}$$

This approximation is made to reduce the computational cost in solving the non-linearity of the inverse scattering problem while the near-field nature of the radar configuration is considered.

To measure only the z–component of the scattered field, the scattering model in (2.5) can be expressed as

$$E_{sc}(r_e, \omega) = -j\omega\mu_{sp}k_p^2 \int_\Omega (G_{ex}^{xz}G_{in}^{zx} + G_{ex}^{yz}G_{in}^{zy} + G_{ex}^{zz}G_{in}^{zz})\chi(r)dr \qquad (2.8)$$

with the dyadic Green's function being the proper sum of the pairwise combination of the vector components [Per14; Dev12]. The reconstruction of the contrast function by inversion of (2.8) still poses mathematical difficulties due to a severely ill-posed inverse problem [BB98]. To recover the well-posedness of the problem and to achieve a generalized stable solution, some regularization strategies are typically employed. However, these strategies can be time-consuming, especially when large data sets are employed. In order to maintain low computational cost, we retrieve the unknown contrast function χ by discretizing equation (2.8) with the method of moments and exploiting the adjoint operator [Dev12]. Thus, the discretized version of the unknown contrast function in the imaging area Ω can be obtained as

$$x = \mathbf{L}\,\mathbf{e}_{sc} \qquad (2.9)$$

where $\mathbf{L}$ is the adjoint discretized operator for (2.8) and $\mathbf{e}_{sc}$ is the samples of the measured scattered field. The spatial mapping of the discretized contrast function x is the tomographic image of the interrogation area.

2.4.3 Measurement Integration

In the process of image formation, multiple measurements from different array positions are coherently integrated to suppress rough surface clutter. At each array position, the view angle of the ROI varies with a maximum difference between the first and last array positions along the x–direction. Instead of integrating measurements from consecutive array positions as performed in [LDS14], 16 different array positions are chosen, which corresponds to eight full aperture measurements with each pair separated by four array positions as indicated with black parallel lines in Fig 2.5.

A tomographic image is composed of four image segments, each of dimension 4 m × 10 m. Fig 2.5 indicates the eight full aperture measurements between −19.33 and −5 m used to image the first segment (dashed rectangle) corresponding to the farthest viewpoint. The full aperture measurements used to construct the second, third, and fourth image segments range from −15.33 to −1 m, −11.33 to 3 m, and −7.33 to 7 m, respectively. These sets of full aperture measurements are illustrated in Appendix A. A tomographic image is thus constructed by integrating 32 full aperture measurements from x = −19.33 m to x = 7 m. Translating the set of measurements by one array position, another image is generated, providing a different viewpoint of the interrogation area. Compared to the image in Fig. 2.6 corresponding to the farthest viewpoint, the tomographic image from the viewpoint closest to the scene

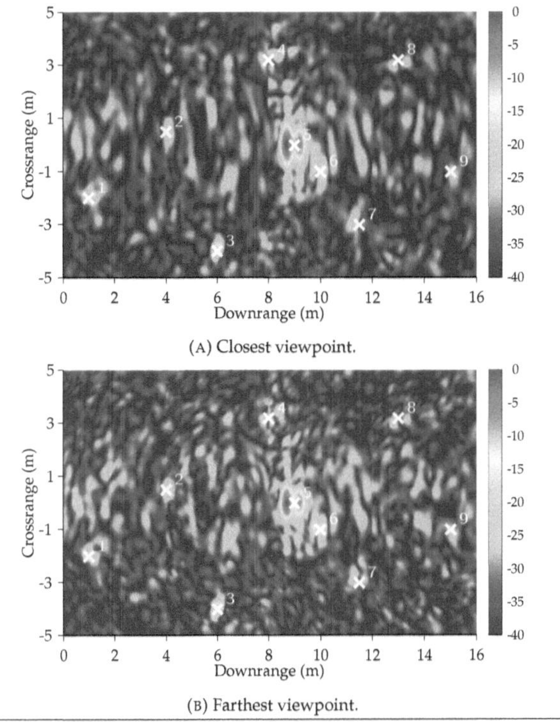

(A) Closest viewpoint.

(B) Farthest viewpoint.

FIGURE 2.6: Tomographic images for low surface roughness ($h_{rms} = 0.8\,\mathrm{cm}$, $l_c = 14.26\,\mathrm{cm}$). White crosses indicate targets.

is realized by integrating measurements from $x = -16\,\mathrm{m}$ to $x = 10.33\,\mathrm{m}$. By this means, a total image of $M = 11$ providing various views of the ROI is generated. For more details of the multiview tomographic imaging approach, refer to [Com+17].

The images in Figs. 2.6 and 2.7 are normalized to 40 dB dynamic range, and each consists of $N_x = 1153$ pixels in downrange and $N_y = 721$ pixels in crossrange with a resolution of 5 cm. The targets located near the boundaries of the investigation area are less visible since they are outside of the mainlobe of the antenna array whose physical aperture is smaller than the image extent in cross range. Compared to the landmines placed on the surface, the buried landmines are more challenging to discern due to the clutter caused by the radar back-scatter from the rough ground surface. In general, recognizing a plastic landmine is harder than a metallic one, since its return signal energy is comparable to that of clutter. The difference in clutter returns across the different viewpoints is clearly visible and illustrates the non-stationary behavior of the scattering from the rough ground surface. Consistent dual signatures in close proximity result from the metallic mines that reside along the radar platform trajectory in low surface roughness, such as targets 5 and 6 in Fig. 2.5. As

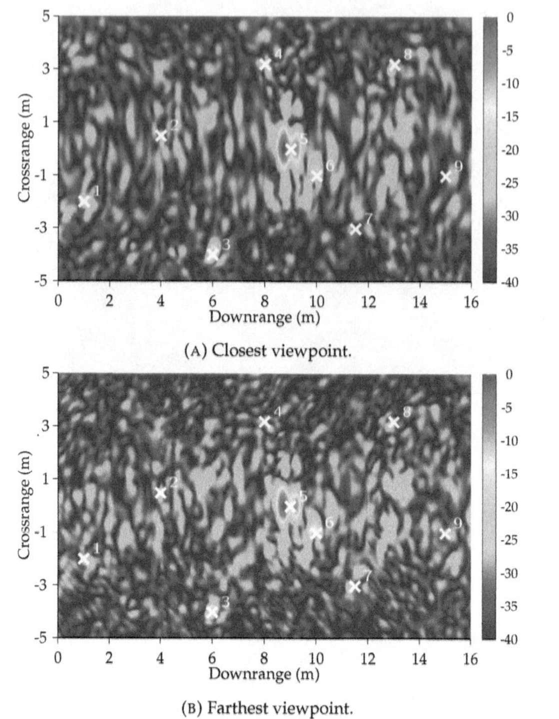

(A) Closest viewpoint.

(B) Farthest viewpoint.

FIGURE 2.7: Tomographic images for high surface roughness ($h_{rms} = 1.6\,\text{cm}$, $l_c = 14.93\,\text{cm}$). White crosses indicate targets.

expected, the higher surface roughness leads to stronger clutter echoes, thus making the detection more challenging.

Chapter 3

Non-Robust Detection Techniques

In this chapter, non-robust detection techniques for landmine detection with FL-GPR are considered. The aim is, given a set of acquired FL-GPR images, to obtain a single binary image providing an indication of the presence or the absence of targets.

3.1 Motivation

Due to interface roughness and subsurface clutter, the detection of targets using the FL-GPR system in a rough surface environment is challenging. In an acquired FL-GPR image, target signatures are typically comparable in strength to the clutter caused by the rough ground surface, which makes discerning targets difficult. Although [SL03; LD12; CCM18] have attempted to overcome the challenge of landmine detection in the FL-GPR image, none of the proposed detection approaches offered a statistically significant performance because they did not consider statistical models of the underlying problem. In this chapter, the detection problem is formulated as a statistical test between two possible hypotheses: the null hypothesis H_0 (mines absent) and the alternative hypothesis H_1 (mines present). An alternate detection approach that is based on the pixel-wise LRT is examined, modeling clutter and target pixels in the FL-GPR image using parametric families of distributions.

The test is applied to radar images obtained from multiple viewpoints of the ROI through multi-view fusion schemes. These techniques offer us a detection performance with a maximum probability of detection under a constraint on the probability of FAs.

3.2 Simple Threshold Detector

Consider M FL-GPR images obtained from multiple viewpoints of the ROI. The detection technique aims to process the series of images such that information about the presence or absence of targets is perceived.

An FL-GPR image is represented as a two-dimensional array of normalized pixel intensities with respect to the maximum intensity value

$$x_m(i,j) \in [0,1], \quad i = 1,\ldots,N_x, \quad j = 1,\ldots,N_y, \quad m = 1,\ldots,M \tag{3.1}$$

where m denotes the viewpoint, i corresponds to the image's pixel index in the down-range and i corresponds to the image's pixel index in the cross-range. An intuitive way to perform detection is by a simple threshold technique. The radar image is binarized by an image threshold ϑ as

$$d_m(i,j) = \begin{cases} 1, & x_m(i,j) > \vartheta \\ 0, & x_m(i,j) \leq \vartheta, \end{cases} \tag{3.2}$$

where each pixel is compared to a predetermined threshold to decide whether a target is present or not at the pixel under test. The series of images can be fused, for instance, through a simple pixel-wise multiplication [AA08] resulting in a binary reference image

$$D(i,j) = \prod_{m=1}^{M} d_m(i,j) \tag{3.3}$$

for M available FL-GPR images for detection. The pixel-wise multiplication can reduce the influence of the ground clutter in the resulting binary fused image $D(i,j)$. Across multiple views of sensing the ROI, it is assumed that the amplitude of the clutter varries significantly, while the same position of targets is shown in all images. From (3.3), a target is declared to be present at pixel $D(i,j)$ if, and only if, strong signals at coordinate (i,j) are observed in all radar images. If strong signals are only observed in one or a few FL-GPR images, they will be mitigated in the multiplication operation and attributed to the clutter.

Considering the image formation from multiple views, a target might appear only in some viewpoints, especially for one at the boundaries of the imaging scene, which may fall outside the radar beam. The pixel-wise multiplication is effective only when the target signatures are apparent in all viewpoints. In the case where the target is visible in almost all views but not in one viewpoint, this will be attributed to clutter, which might lead to a missed target, which is particularly worrisome in landmine detection as it may be potentially deadly. In order to reduce the potential of a missed target, another fusion technique can be applied as follows

$$\sum_{m=1}^{M} d_m(i,j) \underset{0}{\overset{1}{\underset{H_0}{\gtrless}}} M/2. \tag{3.4}$$

This fusion technique will declare a target present at pixel (i,j) in the resulting fused image if the target signatures are significant in more than the half viewpoints, which is similar to a voting strategy with two potential candidates. Compared to the pixel-wise multiplication, reduced potential of a missed target can be obtained by using this fusion strategy, but it might result in more FAs, especially due to strong unwanted back-scattered signals or echoes occuring in the imaging scene close to the radar beam. However, both the fusion strategies in (3.3) and (3.4) require a 'good' chosen threshold ϑ, which is a non-trivial task.

3.3 Likelihood-Ratio Test Detector

An alternative technique to the simple threshold detector is the LRT [Kay98], which provides a minimum missed detection (MD) (type II error) probability under a certain constraint on the FA (type I error) probability (P_{FA}). The detection problem is defined as a pixel-wise test between two possible hypotheses:

$$\begin{aligned} H_0 &: \text{ mines absent at pixel } (i,j) \\ H_1 &: \text{ mines present at pixel } (i,j). \end{aligned} \tag{3.5}$$

Assume the pixel intensities of all M consecutive FL-GPR images to be independently and identically distributed (iid). For the m-th radar image, a pixel-wise LRT is given by

$$L_m(i,j) = \ln\left(\frac{p_1(x_m(i,j))}{p_0(x_m(i,j))} \right) \underset{H_0}{\overset{H_1}{\gtrless}} \gamma, \quad m = 1,\ldots,M \tag{3.6}$$

where $x_m(i,j)$ denotes the pixel intensity at coordinate (i,j) of the m-th radar image, p_0 is the conditional probability density function (pdf) under hypothesis H_0, and p_1 is the conditional pdf under hypothesis H_1. The likelihood ratio L is compared to a threshold γ for a given FA rate constraint $P_{\text{FA}} = \alpha$, which can be determined from

$$P_{\text{FA}} = \Pr(L(x) > \gamma \,|\, H_0) = \int_{\gamma}^{\infty} f_L(l \,|\, H_0)\, dl = \alpha \tag{3.7}$$

where $f_L(l \,|\, H_0)$ denotes the density of the likelihood ratio under the null hypothesis.

In the case of simple hypotheses, a parametric family of distribution models with one or more real parameters is typically involved in the LRT. The pixel values of the FL-GPR image under H_0, which represent clutter or unwanted echoes due to rough ground surface, have been shown to follow the Rayleigh model, i.e., $X \sim \text{Rayleigh}(\sigma_0)$ with σ_0 being the distribution parameter [Roo+99; LDS14]. On the other hand, the target pixels from the FL-GPR image approximately follow a three-component Gaussian mixture model (GMM) [Com+17; Com+18]. That is, $X \sim \sum_{k=1}^{3} \phi_k \mathcal{N}(\mu_k, \sigma_k^2)$, where $\mathcal{N}(\mu_k, \sigma_k^2)$ denotes a Gaussian distribution with mean μ_k and variance σ_k^2, and ϕ_k is the mixture weight. The three Gaussian mixture components have been determined as a valid estimate via extensive numerical simulations, which represent weak, medium and strong target signatures in the FL-GPR image. As the pdfs have very little impact outside the interval of the normalized pixel intensities, $x \notin [0, 1]$, the unknown statistic parameters can be obtained through maximum likelihood estimation (MLE) of the non-truncated Rayleigh and Gaussian pdfs [DAZ09]:

$$p_0(x) = \frac{x}{\sigma_0^2} e^{\frac{-x}{2\sigma_0^2}}, \quad p_1(x) = \sum_{k=1}^{3} \phi_k \mathcal{N}(x \mid \mu_k, \sigma_k^2). \tag{3.8}$$

The maximum likelihood of the distribution parameter under the null hypothesis σ_0, which can be estimated using the clutter pixels $x_c(i, j) = \{x \in [0, 1] \mid H_0\}$ obtained from a target-free FL-GPR image, is given by

$$\hat{\sigma}_0 = \sqrt{\frac{1}{2N_c} \sum_{i=1}^{N_{xc}} \sum_{j=1}^{N_{yc}} x_c^2(i, j)} \tag{3.9}$$

where N_{xc} is the number of clutter pixels in the downrange, N_{yc} is the number of clutter pixels in the cross-range and $N_c = N_{xc} \cdot N_{yc}$ is the total number of pixels. The parameters $\{\phi_k, \mu_k, \sigma_k\}$ of the GMM can be estimated through the expectation-maximization (EM) algorithm [RW84; Moo96; Sen+12] using the target pixels obtained from a clutter-free FL-GPR image.

3.3.1 Multi-View Fusion Schemes

For detecting targets with multiple images, we can apply three different fusion schemes. First, likelihood ratios from each single-view image can be fused before comparing them with a decision threshold. Other fusion schemes can be used by applying the LRT detector to individual images. The resulting binary images are then fused using pixel-wise multiplication or the voting strategy.

Consider a set of images $\{x_m(i, j), i = 1, \ldots, N_x, j = 1, \ldots, N_y, m = 1, \ldots, M\}$. Under the iid assumption of pixel intensities from the consecutive viewpoints, the LRT detector applied to the M images is given by

$$L(i, j) = \sum_{m=1}^{M} \ln \left(\frac{p_1(x_m(i, j))}{p_0(x_m(i, j))} \right) \underset{H_0}{\overset{H_1}{\gtrless}} \gamma. \tag{3.10}$$

The likelihood ratio L is compared to a threshold γ defined in (3.7) for a given false alarm rate constraint. A fused binary image F_A is obtained as

$$F_A(i,j) = \begin{cases} 1, & L(i,j) > \gamma \\ 0, & L(i,j) \leq \gamma. \end{cases} \tag{3.11}$$

The parameters of the clutter and target statistics are estimated by considering all M images under H_0 and H_1, respectively. The parameter statistics of the single-view image are applicable for the multi-view case, and the MLE is used to estimate the parameters of the corresponding Rayleigh model and GMM.

Alternate fusion schemes that apply the LRT detector to the individual images are considered. For the m-th FL-GPR image, a single-view decision is obtained as

$$\delta_m(i,j) = \begin{cases} 1, & L_m(i,j) > \gamma \\ 0, & L_m(i,j) \leq \gamma \end{cases} \tag{3.12}$$

with the likelihood ratio L_m and the threshold γ defined in (3.6) and (3.7), respectively. Through a pixel-by-pixel multiplication, a fused binary image F_B can be obtained as

$$F_B(i,j) = \prod_{m=1}^{M} \delta_m(i,j), \quad i = 1,\ldots,N_x, \, j = 1,\ldots,N_y. \tag{3.13}$$

This fusion scheme can probably reduce a high number of false alarms because only test statistics that are significant in all viewpoints will be attributed to the presence of a target. However, this is more susceptible to MD since the target presence will fail to declare even if only one of the multi-view test statistics is insignificant. To avoid the shortcoming of the pixel-wise multiplication, a fusion scheme based on a voting strategy F_C can be applied as

$$F_C(i,j) = \begin{cases} 1, & \delta(i,j) > M/2 \\ 0, & \delta(i,j) \leq M/2 \end{cases} \quad \text{for} \quad \delta(i,j) = \sum_{m=1}^{M} \delta_m(i,j). \tag{3.14}$$

In the fusion output, a target presence will be declared at a pixel that has significant test statistics in more than half of the multi-view images. If a test statistic is significant in only less than or equal to half of the total viewpoints, it will be attributed to clutter. This fusion strategy offers reduced FAs with a lower potential of MD compared to the pixel-wise multiplication strategy in (3.13).

3.3.2 Multiple Testing Correction

When performing pixel-wise LRT, the problem of multiple hypothesis testing (MHT) may arise; that is, the more pixels that are tested, the higher the probability of obtaining type I errors (FAs). The pixel-wise test, which is designed to control the probability of FA at a certain level ($P_{\text{FA}} = \alpha$), should, therefore, be corrected to cope with this problem. In hypothesis testing in FL-GPR, often a very low α is chosen to avoid a large number of FAs [Kos+02; JLZ12; CCM18]. A lowered α indeed leads to a reduction in the number of FAs, but it is a questionable error measure in this case. In order to provide reliable detection with pixel-wise LRT, several alternative error measures to MHT namely the family-wise error rate (FWER) and the false discovery rate (FDR), are considered.

The FWER [HT87], a commonly used type I error measure in MHT, is defined as the probability of obtaining at least one FA when performing N tests:

$$\text{FWER} = \Pr(N_{\text{FA}} \geq 1) = 1 - (1 - \alpha)^N \tag{3.15}$$

where N_{FA} is the number of FAs. The FWER relates to a single hypothesis test at level α exponentially with degree N, as in (3.15). At a certain large value of N, it will be almost sure to commit at least one error in the detection when performing with single P_{FA}-based threshold procedures. A popular procedure to control FWER is the Bonferroni correction (BC) method [HT87], which adjusts the single threshold α such that the FWER is obtained at the desired level α_{w} defined as

$$\alpha_{\text{w}} = \alpha / N_{\text{xy}}, \quad N_{\text{xy}} = N_{\text{x}} \cdot N_{\text{y}}. \tag{3.16}$$

In order to obtain FWER at the desired level α_{w}, the individual pixel-wise LRT as in (3.6) has to be performed at a threshold for a given FA rate constraint α / N_{xy}. The higher the total number of tested pixels N_{xy}, the lower the test threshold γ would be. This very tight threshold can result in a high MD rate of the FWER control procedures, which is too risky for landmine detection.

An alternative procedure to control FWER is the Holm–Bonferroni (HB) method, which is, in general, more powerful than BC [AG96]. The HB procedure sorts p-values from the lowest to the highest $p_1 \leq \cdots \leq p_{N_{\text{xy}}}$ and compares them to a series of levels $\alpha_{\text{w}} / N_{\text{xy}}, \alpha_{\text{w}} / (N_{\text{xy}} - 1), \ldots, \alpha_{\text{w}} / 2, \alpha_{\text{w}}$, respectively. A p_n-value is the probability of obtaining likelihood ratio L at least as extreme as (greater than or equal to) the actual observed value l, under the assumption that the hypothesis H_0 is true, hence

$$p_n(l) = \Pr(L \geq l \mid H_0), \quad n = 1, \ldots, N_{\text{xy}} \tag{3.17}$$

where $N_{\text{xy}} = N_{\text{x}} \cdot N_{\text{y}}$ is the number of hypothesis tests (pixels). The procedure decides in favor of H_1 (target presence) for all $p_1, \ldots, p_{n-1}$ if $p_n \leq \alpha_{\text{w}} / (N_{\text{xy}} - n + 1)$. This method controls the FWER to be no higher than a certain level α_{w} with an offer of more detection power than the BC procedure.

The FDR is another type I error measure used practically in MHT [BH95], which is the expected ratio of the number of FAs to the total number of discoveries (rejections of the null hypothesis)

$$\text{FDR} = \text{E}\left[\frac{N_{\text{FA}}}{N_{\text{D}}}\right] = \text{E}\left[\frac{N_{\text{FA}}}{N_{\text{FA}} + N_{\text{CD}}}\right] \tag{3.18}$$

where $N_{\text{D}} = N_{\text{FA}} + N_{\text{CD}}$ is the number of discoveries, N_{FA} is the number of FAs (decide H_1, H_0 true) and N_{CD} is the number of correct detections (decide H_1, H_1 true). A common procedure to control the FDR at a certain level α_r is the Benjamini–Hochberg (BH) method, which sorts the N p-values in ascending order $p_1 \leq \cdots \leq p_{N_{\text{xy}}}$, selects the rejection index

$$n^* = \left\{\max n \in 1, \ldots, N_{\text{xy}} \mid p_n \leq \frac{n}{N_{\text{xy}}}\alpha_r\right\} \tag{3.19}$$

and then declares target presence (hypothesis H_1) for all $p_1, \ldots, p_{n^*}$. The BH procedure controls FDR to not be higher than the desired level α_r. Since FDR allows an arbitrary N_{FA} if N_{D} is large enough, a procedure controlling FDR, in general, tends to have more detection power than the one controlling the FWER. Loosely speaking, if we end up with a larger number of discoveries, we can afford a larger absolute number of FAs.

3.4 Numerical Results

In order to examine the performance of the detection techniques, the problem of detecting landmines in a rough surface environment from the scenario described in Section 2.4 is considered. Single and multiple FL-GPR images obtained from different viewpoints of the ROI are taken into account for the detection.

3.4.1 Simple Thresholding

A simple threshold detector is applied to the normalized single viewpoint FL-GPR image in Fig. 2.6(a). Each image pixel is compared to a pre-determined threshold to decide whether or not the target is present at the pixel under test as given in (3.2). Fig. 3.1(a) shows the results in a binary image for a threshold value of $-20\,\text{dB}$, where red indicates the detected target region and black represents the FA. Five targets $\{2, 3, 4, 5, 6\}$ are detected, but four targets $\{1, 7, 8, 9\}$ are missed. Figs. 3.1(b) and 3.1(c) show the binary images obtained by applying a fusion rule using the pixelwise multiplication as given in (3.3) and the majority voting as given in (3.4) of $M = 11$ multi-view FL-GPR images. As seen, the fusion procedure with the pixelwise multiplication contributes one additional MD (target 4), while the results from the fusion scheme with the majority voting are similar to those from the single view image, with five detected targets and four MDs. Apparently, the use of a threshold of $-20\,\text{dB}$ removes desirable targets as well as the high influence of unwanted clutter.

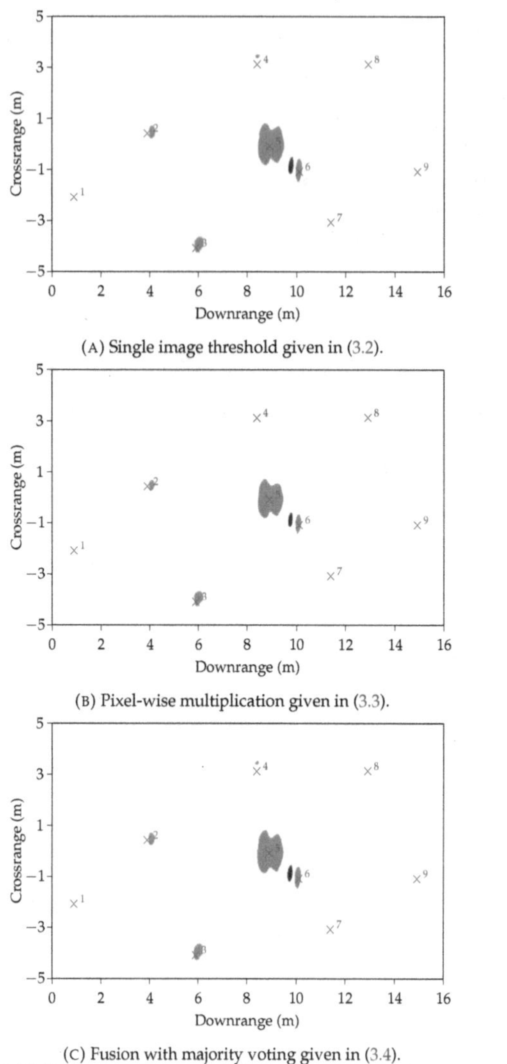

(A) Single image threshold given in (3.2).

(B) Pixel-wise multiplication given in (3.3).

(C) Fusion with majority voting given in (3.4).

FIGURE 3.1: Detection results from simple threshold detectors.

Setting the threshold to a lower value, for example $-30\,\mathrm{dB}$, tends to preserve a significant amount of clutter, which leads to a high FAR [Com+17]. Choosing a suitable threshold is the main issue of this simple detector. An LRT detector uses a meaningful way of choosing a decision threshold by incorporating statistical behaviours of the underlying detection problem and is described in the following section.

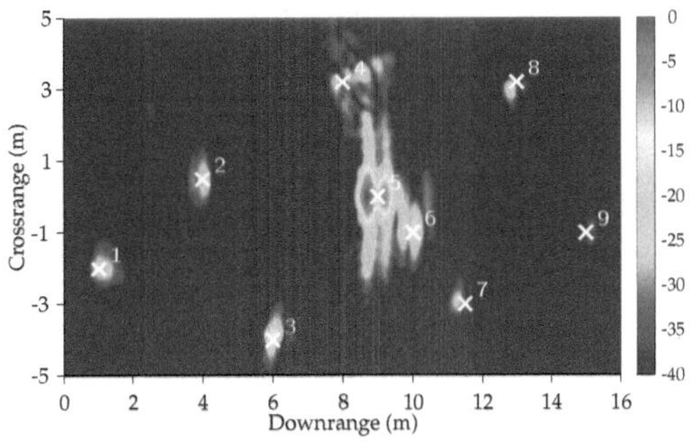

FIGURE 3.2: Training image T, normalized FL-GPR image of the closest viewpoint on flat ground surface.

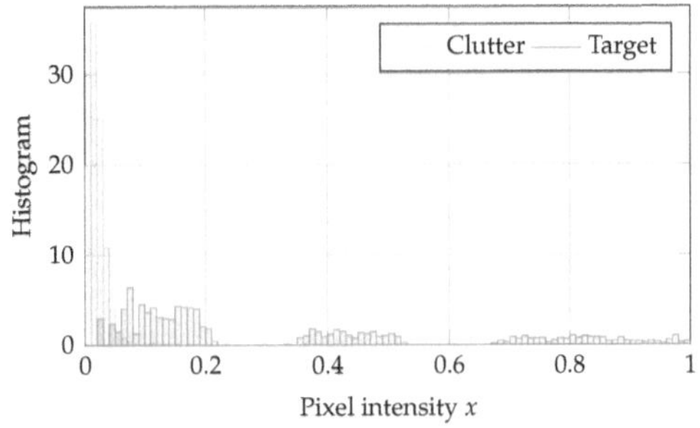

FIGURE 3.3: Histogram of the training data sets of clutter pixel ($\mathcal{X}_c$) and target pixel ($\mathcal{X}_t$).

3.4.2 Training Data Sets

In order to design the LRT detector, *a priori* knowledge about the distribution of the pixel intensities under hypothesis H_0 (mines absent) and hypothesis H_1 (mines present) is needed. In order to obtain this knowledge, two training images, T and C, generated by the NAFDTD software, are used.

Training image T is a clutter-free image as shown in Fig. 3.2, which is a normalized FL-GPR image that was generated with the same simulation parameters as described in Section 2.4, except that the ground was assumed to be flat. A region growing algorithm [AB94; RMK07] is then employed to locate the target pixel regions. First, the center points of the nine target positions are chosen manually. The algorithm then successively adds adjacent pixels to the target region based on the connection criterion $\varphi \pm 3\,\text{dB}$, where φ denotes the pixel intensity of the seed points; see [AB94; RMK07] for more details. The output of the region growing algorithm is a binary

TABLE 3.1: Parameter estimates for p_0, p_1.

$\hat{\sigma}_0$	$\hat{\sigma}_1$	$\hat{\sigma}_2$	$\hat{\sigma}_3$	$\hat{\phi}_1$
0.025	0.050	0.048	0.088	0.601

$\hat{\phi}_2$	$\hat{\phi}_3$	$\hat{\mu}_1$	$\hat{\mu}_2$	$\hat{\mu}_3$
0.212	0.186	0.117	0.430	0.833

mask given by

$$\text{BM}(i,j) = \begin{cases} 1, & \text{target region at } (i,j) \\ 0, & \text{non-target region at } (i,j) \end{cases} \tag{3.20}$$

with $i = 1,\ldots,N_x$ and $j = 1,\ldots,N_y$. By multiplying the original image T with the mask BM, the image $T = T \cdot \text{BM}$ is obtained, which only contains the target pixels and zeros. The training image C is the closest viewpoint FL-GPR image for the low surface roughness profile ($h_{\text{rms}} = 0.8\,\text{cm}$, $l_c = 14.26\,\text{cm}$) as shown in Fig. 2.6(a). The image C is multiplied by the mask image complement BM^c to obtain the image C, which is composed of only clutter pixels and zeros. In this way, we obtain two data sets, $\mathcal{X}_t$ and $\mathcal{X}_c$, containing only target and clutter pixels, respectively:

$$\mathcal{X}_t = \{\, x \mid x \in T,\, x > 0 \,\}, \quad \mathcal{X}_c = \{\, x \mid x \in C,\, x > 0 \,\}. \tag{3.21}$$

The data set $\mathcal{X}_t$ consists of $N_t = 3206$ pixels, while the total pixels in set $\mathcal{X}_c$ equals $N_c = N_{xy} - N_t$, where $N_{xy} = N_x \times N_y$. The histograms of the training data sets $\mathcal{X}_c$ and $\mathcal{X}_t$ are shown in Fig. 3.3. As mentioned in Section 3.3, the authors of [Com+17; DAZ09; LD12] showed that the clutter pixels are approximately Rayleigh distributed, while the distributions of the target pixels can be approximated by a Gaussian mixture distribution. Using the training data detailed above, maximum likelihood estimates for the parameters of the densities in (3.8) are obtained as shown in Table 3.1.

3.4.3 Likelihood-Ratio Test

Having estimated the parameter statistics of the densities (p_0, p_1) under both hypotheses, the LRT detector is applied to the single-view and multi-view FL-GPR images. A binary image as shown in Fig. 3.4 is obtained by applying the LRT detector to a single-view FL-GPR image for a nominal false alarm rate of $\alpha = 0.05$. As seen, all nine targets are successfully detected, but there exist a high number of FAs as well. The large red area at around $9\,\text{m}$ downrange is because of the strong reflections from the metallic AT landmine placed on the surface. In addition, due to their strong reflections and close proximity, targets 5 and 6 emerge as a single region.

Fig. 3.5 depicts the results of the LRT detector applied to multi-view FL-GPR images via three different fusion schemes (F_A, F_B and F_C) as given in (3.11), (3.13) and (3.14), respectively. Note the set of image statistics for the single-view case is also

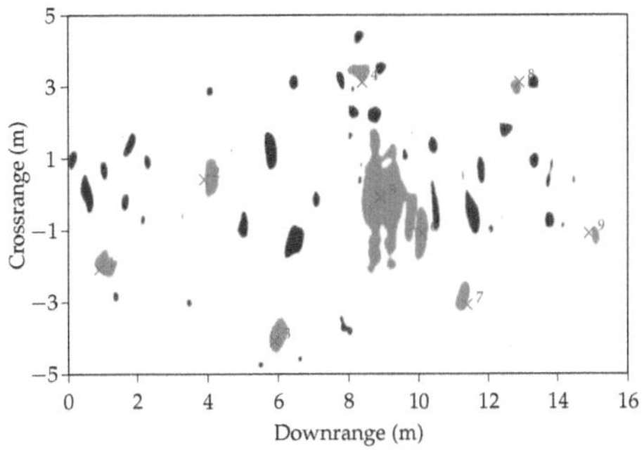

FIGURE 3.4: Detection results of LRT detector applied to a single-view FL-GPR image. Color coding: detected targets (red), FAs (black).

applicable for the multi-view case. Fig. 3.5(a) shows the binary image F_A obtained by applying the fusion scheme in (3.11) to $M = 11$ radar images, which assumes the images from the various viewpoints to be iid under both hypotheses. The fusion is performed through multiplication of likelihood ratios calculated from every single-view image. Compared to the single-view result of Fig. 3.4, a slight increase in performance is achieved. However, the existence of a prominent number of FAs causes the performance to be unsatisfactory. The lack of performance improvement compared to the single-view case can be attributed to the invalidity of the independence assumption across all M images due to the relatively small increment of array positions during multi-view operation. Fig. 3.5(b) depicts the binary image F_B obtained through pixel-wise multiplication as given in (3.13). The LRT detector is first performed to every single-view images; then the binary image is obtained through pixel-by-pixel multiplication. Compared to the result of Fig. 3.5, a high reduction in the number of FAs is achieved and all targets can still be detected. Note that this fusion rule has more potential for causing MDs since a target presence will not be declared even if only one of the multi-view tests is not significant. To prevent this shortcoming, a fusion rule based on majority voting, whose results are shown in Fig. 3.5(c), is examined. However, a considerably higher number of FAs is obtained using this fusion scheme even compared to the single-view case. Finally, the problem of multiple testing correction as discussed in Section 3.3.2 is examined. Fig. 3.6 depicts the corresponding result with the FDR controlled at level of $\alpha_r = 0.05$. Only one target with the strongest reflections placed in the middle of the ROI can be detected. This lack of detection power can be caused by the use of simple error control that assumes the test statistics to be independent across all $N_x \times N_y$ pixels in the radar image. This assumption is not satisfied because there exist homogeneous regions in the image due to to either prominent clutter or salient target signatures. As a first step in this direction, an approach was proposed in [Pam+20a].

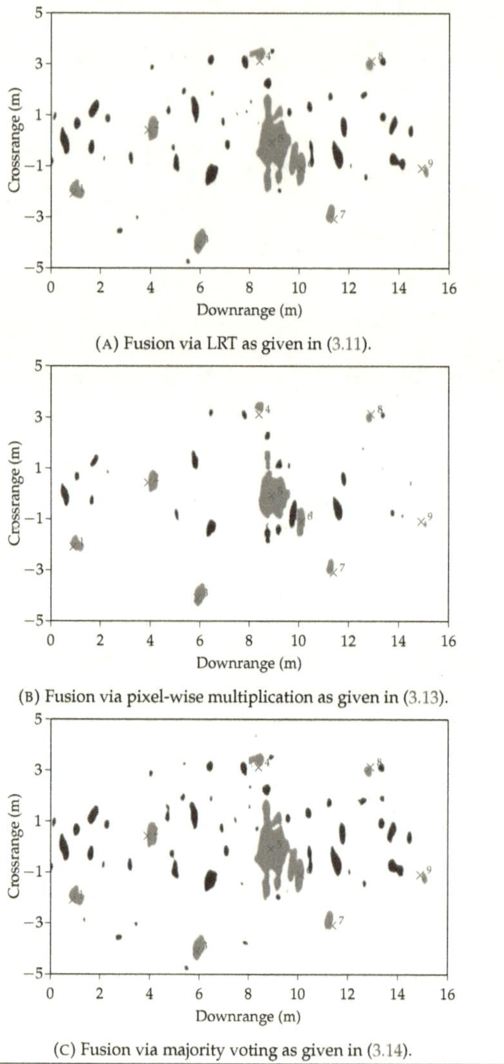

(A) Fusion via LRT as given in (3.11).

(B) Fusion via pixel-wise multiplication as given in (3.13).

(C) Fusion via majority voting as given in (3.14).

FIGURE 3.5: Detection results of LRT detector applied to multi-view FL-GPR images. Color coding: detected targets (red), FAs (black).

3.5 Summary

In this chapter, the problems of landmine detection with FL-GPR in a rough surface environment and non-robust detection techniques are considered. Detectors based on a simple threshold scheme and an LRT are presented. The detectors are applied to tomographic FL-GPR images obtained from multiple viewpoints of the ROI. A

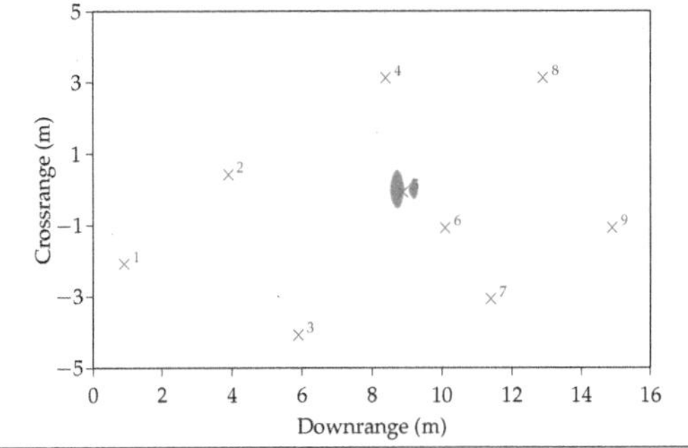

FIGURE 3.6: Detection results with FDR control. Color coding: detected targets (red), FAs (black).

parametric family of distributions is employed for the LRT detector, providing the statistics on the FL-GPR image. Three multi-view image fusion procedures based on the LRT are considered to obtain a desired performance of a reduced number of FAs and an increased detection power. The performance of the detectors is evaluated using numerical radar data of shallow buried targets. The superior performance of the multi-view approach over the singe-view imaging case is demonstrated through its ability to reduce the influence of clutter on the detection.

Chapter 4

Robust Detection Techniques

In this chapter, robust techniques for landmine detection in FL-GPR imagery are considered. The main objective is to design a robust LRT detector that maximizes the worst-case performance over all feasible image clutter and target distribution pairs and, hence, eliminates the need for strong assumptions about the statistics of the radar image. The content presented in this chapter is partly taken from [Pam+20b; Pam+19; Pam+18].

4.1 Motivation

The detection techniques based on the pixel-wise LRT described in Chapter 3 do not provide us with a performance guarantee in case of deviations of the statistical models [Zou+18]. The assumption that the distribution of the FL-GPR image follows a particular parametric model under each hypothesis is too strong for its practical application in landmine detection. In FL-GPR operation, the distributions of targets and clutter may deviate from the assumed models due to the varying nature of the

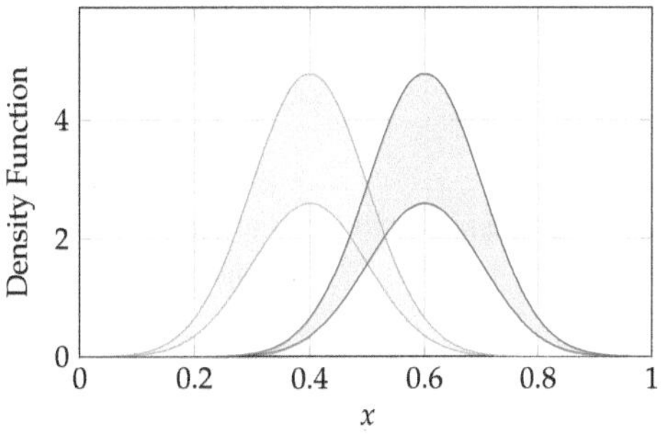

FIGURE 4.1: Illustration of two feasible density bands.

interrogated environment [LD12; Kos+02]. This chapter introduces an uncertainty distribution model for the FL-GPR image. Instead of modeling the distribution with a parametric family, we construct two density bands within which the corresponding pdfs of feasible distributions are assumed to lie. The detector is then designed such that it minimizes the maximum error probability for all possible density pairs within the two bands. The density band model was initially introduced by Kassam [Kas81]. In Fig. 4.1 the density band principle is illustrated. The reason behind following a minimax approach is that accurate estimation of the distribution of the background clutter, given its non-stationary behavior, is highly challenging. The proposed technique overcomes this issue since it does not need an accurate estimate in the first place and is guaranteed to perform well over a set of feasible distributions. In [Kas81; FZ16; Pam+20b; FZP21], it is shown that under very mild assumptions a minimax optimal test for such an uncertainty model is guaranteed to exist.

4.2 Image-Domain Statistical Model

In this section, an image-domain statistical model is provided, which forms the basis for the robust detection techniques. The model assumes a ground-based FL-GPR imaging system as described in Chapter 2.

Let $x \in [0,1]$ be the normalized pixel intensity of an FL-GPR image. As described in Section 3.3, the clutter pixels representing the interactions with the rough ground surface have been shown to exhibit Rayleigh statistics [Roo+99; LDS14], whereas the target pixels from the FL-GPR image approximately follow a three-component Gaussian mixture model [Com+17]. The three Gaussian mixture components represent weak, medium and strong target signatures in the radar image. In practice, the distributions of the pixel intensities are not known a priori. Replacing the unknown statistical parameters with their maximum likelihood estimates as shown in

Table 3.1, the corresponding pdfs are given by

$$\hat{p}_0(x) = \frac{x}{\hat{\sigma}_0^2}\, e^{\frac{-x}{2\hat{\sigma}_0^2}}, \quad \hat{p}_1(x) = \sum_{k=1}^{3} \hat{\phi}_k\, \mathcal{N}(x \mid \hat{\mu}_k, \hat{\sigma}_k^2) \tag{4.1}$$

where $\hat{\sigma}_0^2$ is the parameter estimate of the Rayleigh distribution. The estimates of the mean and the variance of k-th Gaussian component are denoted by $\hat{\mu}_k$ and $\hat{\sigma}_k^2$ and the weight estimate is denoted by $\hat{\phi}_k$.

Due to variability in the characteristics of the ground medium, uncertainty in the aforementioned parametric distribution models is likely to occur [SL03; Kos+02; LD12]. Such uncertainty can be accounted for by constructing a confidence band around density estimates. The bootstrap resampling and kernel density estimation methods [ZI04; Pam+19] are considered to construct a density band. Assuming availability of N clutter pixels and N target pixels constituting clutter sample $X_0 = \{x_{0,n}, n = 1, \ldots, N\}$ and target sample $X_1 = \{x_{1,n}, n = 1, \ldots, N\}$ respectively, bootstrap resamples $\{X_{0,b}^*, b = 1, \ldots, B\}$ and $\{X_{1,b}^*, b = 1, \ldots, B\}$ are generated. A density estimate is obtained by applying a Gaussian kernel density estimator to the bootstrap resample as

$$\hat{p}_{0,b}(x) = \frac{1}{Nh} \sum_{n=1}^{N} \varphi\left(\frac{x - x_{0,n}^*}{h}\right), \quad \forall x_{0,n}^* \in X_{0,b}^*$$

$$\hat{p}_{1,b}(x) = \frac{1}{Nh} \sum_{n=1}^{N} \varphi\left(\frac{x - x_{1,n}^*}{h}\right), \quad \forall x_{1,n}^* \in X_{1,b}^* \tag{4.2}$$

where φ is the pdf of the standard normal distribution and h is a smoothing parameter. A confidence interval consists of a lower bound and an upper bound, which are determined as point-wise minimum and maximum of all kernel-based density estimates:

$$\underline{p}_0(x) = \min\{\,\hat{p}_{0,1}(x), \hat{p}_{0,2}(x), \ldots, \hat{p}_{0,B}(x)\,\}$$

$$\underline{p}_1(x) = \min\{\,\hat{p}_{1,1}(x), \hat{p}_{1,2}(x), \ldots, \hat{p}_{1,B}(x)\,\} \tag{4.3}$$

$$\bar{p}_0(x) = \max\{\,\hat{p}_{0,1}(x), \hat{p}_{0,2}(x), \ldots, \hat{p}_{0,B}(x)\,\}$$

$$\bar{p}_1(x) = \max\{\,\hat{p}_{1,1}(x), \hat{p}_{1,2}(x), \ldots, \hat{p}_{1,B}(x)\,\} \tag{4.4}$$

The corresponding confidence bands satisfy the following integration condition:

$$\xi_s = \int \bar{p}_s(x) - \underline{p}_s(x)\, dx \geq 0, \quad s \in \{0, 1\} \tag{4.5}$$

which accounts for the uncertainty level in case of deviations from the assumed distribution. As observed in [Pam+20b; Pam+19], the uncertainty $\xi_s \gg 0$ may occur in the FL-GPR image, notably in the range of weak mine signatures and in the mode of the clutter distribution.

4.3   Robust Likelihood-Ratio Test

The idea of robust detection techniques is to design an LRT detector such that it works well under all feasible clutter and target distributions of the FL-GPR image. In contrast to the generalized LRT discussed in Chapter 3, which tries to *reduce* the uncertainty by estimating unknown parameters of the assumed parametric distribution models, a robust LRT detector *tolerates* uncertainty by assuming that the distributions may deviate from the assumed models.

Consider M FL-GPR images, each of size $N_x \times N_y$ pixels, which are generated ranging from the farthest to the closest viewpoints of the ROI along the vehicle trajectory. The pixel intensities of all M radar images are assumed to be identically distributed. For the m-th viewpoint, the detection problem is defined as a pixel-wise test between two possible hypotheses, namely, the null hypothesis H_0 (mines absent) and the alternative hypothesis H_1 (mines present). Let $x_m(i,j), i = 1, \ldots, N_x, j = 1, \ldots, N_y$ denote the normalized intensity of the (i,j)-th pixel corresponding to the m-th viewpoint. Then, a pixel-wise LRT can be expressed as in (3.6). However, this test requires the distributions to be known exactly.

With robust hypothesis testing, the pixel distribution of the FL-GPR image under hypothesis H_s is assumed to be an element of an uncertainty set denoted by $\mathcal{P}_s, s = \{0, 1\}$. The binary hypotheses are, therefore, transformed into composite hypotheses

$$
\begin{aligned}
H_0 &: P_m \in \mathcal{P}_0 \\
H_1 &: P_m \in \mathcal{P}_1
\end{aligned}
\tag{4.6}
$$

for all $m \in \{1, \ldots, M\}$, and P_m denotes the distribution of x_m. The uncertainty sets $\mathcal{P}_0$ and $\mathcal{P}_1$ are chosen such that they adequately capture the distributional uncertainties of the underlying detection problem. Here, we use Kassam's density band model for this purpose, which is a generalization of the ϵ-contamination and has been shown to provide a decent compromise between flexibility and tractability [Kas81; FZ16; FZP21]. The sets of feasible distributions are characterized as

$$
\begin{aligned}
\mathcal{P}_0 &= \left\{ p_0 \mid \underline{p}_0(x) \leq p_0(x) \leq \bar{p}_0(x), \int p_0(x)\, \mathrm{d}x = 1 \right\} \\
\mathcal{P}_1 &= \left\{ p_1 \mid \underline{p}_1(x) \leq p_1(x) \leq \bar{p}_1(x), \int p_1(x)\, \mathrm{d}x = 1 \right\}
\end{aligned}
\tag{4.7}
$$

where $\underline{p}_s$ and $\bar{p}_s$ denote lower and upper bounds on the true density, respectively. The bounds need to be non-negative functions satisfying

$$
\int \underline{p}_s(x)\, \mathrm{d}x \leq 1 \leq \int \bar{p}_s(x)\, \mathrm{d}x, \quad s \in \{0, 1\}
\tag{4.8}
$$

but can otherwise be chosen freely by the test designer.

4.3.1 Least Favorable Densities

In principal, a minimax robust hypothesis test is designed by finding a pair of densities $(g_0, g_1) \in \mathcal{P}_0 \times \mathcal{P}_1$ which is *least favorable* in the sense that it simultaneously maximizes both error probabilities (type I and type II) among all feasible densities. If such a pair exists, the corresponding minimax optimal test can be shown to be a threshold test whose test statistic is the likelihood ratio of the least favorable density (LFD) pair [Hub65; FZ16; FZP21]. When using the density band uncertainty model, the LFD can be characterized by [FZ16]

$$g_0(x) = \min\{ \bar{p}_0(x), \max\{ a_0\, g_1(x), \underline{p}_0(x) \}\}$$
$$g_1(x) = \min\{ \bar{p}_1(x), \max\{ a_1\, g_0(x), \underline{p}_1(x) \}\} \tag{4.9}$$

where the constants (a_0, a_1) have to be calculated such that the LFDs are valid densities, i.e., (g_0, g_1) integrate to one. Based on the procedure outlined in [FZ16], an iterative algorithm to construct the LFDs is provided in Table 4.1. The notation $\|\cdot\|$ denotes the L^2-norm chosen for the termination criterion of the iterative algorithm. Starting with an initial guess for the LFDs (g_0^0, g_1^0), the algorithm alternately updates g_0^t and g_1^t by finding a root of

$$f_s(a; g) = \int \min\{\bar{p}_s(x), \max\{a\, g(x), \underline{p}_s(x)\}\}\, \mathrm{d}x - 1 \tag{4.10}$$

which, for a given density g, is a non-decreasing function of the scalar a. The iteration is terminated once both densities have converged within a small tolerance ϱ. The likelihood ratio of the LFDs, $g_1(x)/g_0(x)$, can take six possible values, namely

$$\frac{g_1(x)}{g_0(x)} \in \left\{ \frac{\underline{p}_1(x)}{\bar{p}_0(x)}, \frac{\underline{p}_1(x)}{\underline{p}_0(x)}, \frac{\bar{p}_1(x)}{\underline{p}_0(x)}, \frac{\bar{p}_1(x)}{\bar{p}_0(x)}, a_1, \frac{1}{a_0} \right\}. \tag{4.11}$$

The first four values correspond to regions where g_0 or g_1 coincide with either their lower or their upper bound.

Hence, for the two remaining cases it holds that either $\underline{p}_1(x) < g_1(x) < \bar{p}_1(x)$, which by (4.9) implies that $g_1(x) = a_1\, g_0(x) \Rightarrow g_1(x)/g_0(x) = a_1$, or that $\underline{p}_0(x) < g_0(x) < \bar{p}_0(x)$, which implies that $g_0(x) = a_0\, g_1(x) \Rightarrow g_1(x)/g_0(x) = 1/a_0$.

An interesting special case of the band model can be obtained by relaxing the upper density bounds $(\bar{p}_0, \bar{p}_1 \to \infty)$, so that the LFDs in (4.9) reduce to

$$g_0(x) = \max \{a_0\, g_1(x), \underline{p}_0(x)\}$$
$$g_1(x) = \max \{a_1\, g_0(x), \bar{p}_1(x)\} \tag{4.12}$$

and the possible values of the likelihood ratio become

$$\frac{g_1(x)}{g_0(x)} \in \left\{ \frac{\underline{p}_1(x)}{\underline{p}_0(x)}, a_1, \frac{1}{a_0} \right\}. \tag{4.13}$$

TABLE 4.1: Algorithm to construct the LFDs.

Input: $\underline{p}_0,\ \bar{p}_0,\ \underline{p}_1,\ \bar{p}_1,\ g_0^0,\ g_1^0,\ \varrho$

Output: $g_0,\ g_1$

1: **repeat**
2:      $a_0\ \ = \text{root of } f_0(a; g_1^t)$
3:      $g_0^{t+1} = \min\{\bar{p}_0,\ \max\{a_0\, g_0^t,\ \underline{p}_0\}\}$
4:      $a_1\ \ = \text{root of } f_1(a; g_0^{t+1})$
5:      $g_1^{t+1} = \min\{\bar{p}_1,\ \max\{a_1\, g_0^{t+1},\ \underline{p}_1\}\}$
6:      **if** $\|g_0^t - g_0^{t+1}\| < \varrho$ and $\|g_1^t - g_1^{t+1}\| < \varrho$
7:          **return** $g_0^t,\ g_1^t$
8:      $g_0^t = g_0^{t+1}$
9:      $g_1^t = g_1^{t+1}$

That is, the density band model reduces to the ϵ-contamination model, and the corresponding minimax optimal test becomes Huber's clipped LRT, which is robust against outliers [Hub65]. The ϵ-contamination model is a simple and widely-used uncertainty model in detection and estimation when the majority of the data fits the assumed model [Zou+12; CGR16; Zou+18].

4.3.2 Multi-View Image Fusion

The robust hypothesis test is a pixel-wise LRT of the form (3.6), but with the conditional densities (p_0, p_1) replaced by a pair of LFDs (g_0, g_1)

$$L_{\mathrm{R},m}(i,j) = \ln\left(\frac{g_1(x_m(i,j))}{g_0(x_m(i,j))}\right) \underset{H_0}{\overset{H_1}{\gtrless}} \gamma, \quad m = 1, \ldots, M \tag{4.14}$$

where $x_m(i,j)$ denotes the pixel intensity at coordinate (i,j) of th m-th normalized FL-GPR image, the threshold γ is determined using (3.7) based on a nominal FA rate α, $i = 1, \ldots, N_x$, $j = 1, \ldots, N_y$. The detection results are presented in a binary image whose pixels are given by

$$F_m(i,j) = \begin{cases} 1, & L_{\mathrm{R},m}(i,j) > \gamma \\ 0, & L_{\mathrm{R},m}(i,j) \leq \gamma. \end{cases} \tag{4.15}$$

For M acquired FL-GPR images, a fused binary image is obtained through pixel-by-pixel multiplication as

$$F(i,j) = \prod_{m=1}^{M} F_m(i,j), \quad i = 1, \ldots, N_x, j = 1, \ldots, N_y. \tag{4.16}$$

With this fusion approach, we decide mines present in pixels which have significant test statistic $L_{\mathrm{R},m}$ in all viewpoints, thus reducing FAs.

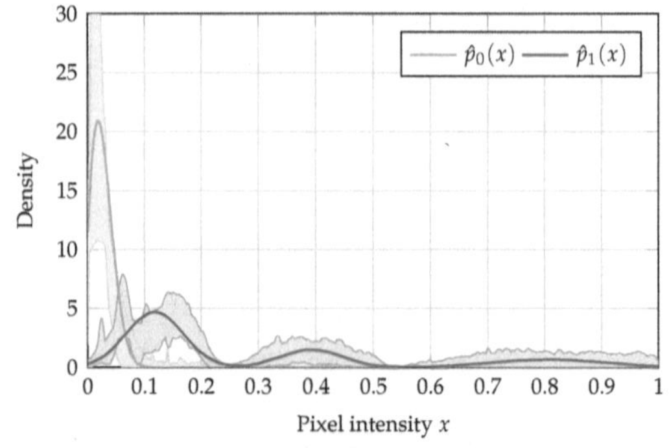

FIGURE 4.2: Parametric density estimates $(\hat{p}_0, \hat{p}_1)$ given in (4.1) with their corresponding confidence bands.

4.4 Construction of Uncertainty Model

In order to calculate the LFDs for the robust LRT, the uncertainty sets $\mathcal{P}_0$ and $\mathcal{P}_1$ have to be constructed. In order to do so, we propose a hybrid approach that combines parametric density estimates constructed from training data with classic nonparametric uncertainty models, which is detailed in this section.

As mentioned in Section 4.2, the authors of [Com+17; DAZ09; LD12] showed that the clutter pixels are approximately Rayleigh distributed, while the distributions of the target pixels can be approximated by a Gaussian mixture distribution. Fig. 4.2 shows the resulting density estimates $(\hat{p}_0, \hat{p}_1)$ of the clutter and target distributions with the confidence bands. The parameters of the distributions are provided in Table 3.1, which are obtained through MLE. The confidence bands are obtained from the original clutter and target samples, as described in Section 4.2. A high distributional uncertainty can be observed from Fig 4.2, especially in the mode of the clutter distribution and in the range of weak mine signatures, i.e., $x < 0.2$. Our aim is to design the robust test such that, on one hand, prior knowledge of the shape of the distributions can still be incorporated, but on the other hand, it is insensitive to deviations from the assumed model. This is accomplished by using the parametric distributions, which are referred to as *nominal distributions*, around which a neighborhood of feasible distributions is constructed. The robust detector is then designed such that it is minimax optimal over the set of feasible distributions. Two ways of choosing the neighborhood are detailed below: the density band model and the ϵ-contamination model.

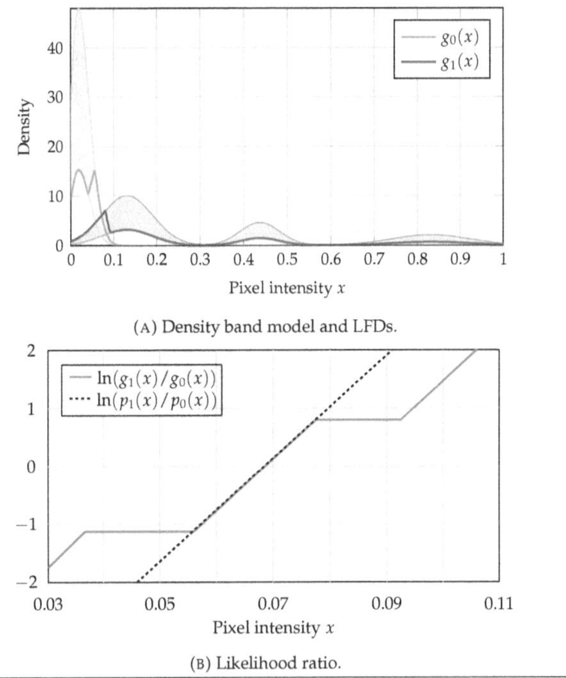

(A) Density band model and LFDs.

(B) Likelihood ratio.

FIGURE 4.3: Density band model in (4.17) with corresponding LFDs and their robust likelihood ratio in logarithmic scale.

4.4.1 Density Band Model

For the first uncertainty model, a band of feasible densities as defined in (4.7) is constructed. The lower and upper density bounds are chosen as

$$\underline{p}_s(x) = 0.8\,\hat{p}_s(x), \quad \bar{p}_s(x) = 2.5\,\hat{p}_s(x), \quad s = 0,1 \tag{4.17}$$

where $\hat{p}_0$ and $\hat{p}_1$ denote the parametric density estimates given in (4.1). By following this uncertainty model, it is assumed that the true distributions are close to the nominal model in the sense that the probability mass on any measurable set $E \subset \mathbb{R}$ is lower bounded by $0.8\,P_s(E)$ and upper bounded by $2.5\,P_s(E)$. This also implies that the deviations from the nominal model are continuous, i.e., events that are highly (un)likely under the nominal distribution are also highly (un)likely under all feasible distributions. The particular values of 0.8 and 2.5 were chosen heuristically based on a series of experiments. More specifically, they have been shown to correspond to a good compromise between guarding against slight but systematic model mismatches and guarding against severe but rare outliers [PFZ18; Pam+18].

With the parametric density estimates of the clutter and target distributions obtained from the training data as described in Chapter 3, the corresponding bands of the feasible densities are shown in Fig. 4.3(a), indicated by the shaded areas. Fig. 4.3(a) also shows the LFDs for this model, i.e., g_0 and g_1. The latter were calculated using the iterative algorithm provided in Table 4.1, which uses the parametric densities p_0 and p_1 as starting points. The tolerance was set to $\varrho = 0.001$. By inspection, each LFD coincides with its upper or lower bound at certain intervals. According to (4.11), it holds that

$$\frac{g_1(x)}{g_0(x)} \in \left\{ \frac{\hat{p}_1(x)}{\hat{p}_0(x)}, \frac{8}{25} \frac{\hat{p}_1(x)}{\hat{p}_0(x)}, \frac{25}{8} \frac{\hat{p}_1(x)}{\hat{p}_0(x)}, a_1, \frac{1}{a_0} \right\} \tag{4.18}$$

which means that the robust likelihood ratio is a scaled version of the nominal likelihood ratio that is clipped at a_1 and $\frac{1}{a_0}$, respectively. Fig. 4.3(b) shows the robust likelihood ratio plotted in a logarithmic scale. The two clipping constants are obtained as $\ln a_1 = -1.125$ and $\ln a_0 = -0.8$ at the intervals $0.0367 \leq x \leq 0.0560$ and $0.0775 \leq x \leq 0.0925$, respectively. In the other ranges, the minimax test statistic is either the same as that of the parametric test (dashed line) or a scaled version.

4.4.2 Outlier Model

For the second uncertainty model, it is assumed that the true distributions are close to the nominal ones in the sense that, under each hypothesis, a majority of the pixels is distributed according to the parametric model, whereas a fraction of ϵ may follow an arbitrary distribution. This is the classic outlier model first studied by Huber [Hub65] and has been used in practice in the past decades [TS08; Zou+18; Mar+18]. For the examples in this section, a contamination ratio of 40 % is assumed, that is

$$\underline{p}_s(x) = 0.6\,\hat{p}_s(x), \quad \bar{p}_s(x) = \infty, \quad s = 0, 1. \tag{4.19}$$

This contamination ratio has been identified as a maximum value such that the LFDs do not overlap completely by constructions, i.e., $g_0(x) \neq g_1(x)$. This amount of contamination was also investigated for the outlier model in [Pam+18]. On the one hand, the assumption that only 60 % of the pixels follow the parametric model is rather pessimistic. On the other hand, it is well in line with the idea of minimax robustness, i.e., preparing for the worst case.

The LFDs for the uncertainty model in (4.19) are shown in Fig. 4.4(a). Note that they are difficult to distinguish by inspection since they are very similar by construction. As stated in (4.13) and shown in Fig. 4.4(b), the corresponding robust likelihood ratio in logarithmic scale satisfies

$$\ln \frac{g_1(x)}{g_0(x)} \in \left\{ \ln \frac{\hat{p}_1(x)}{\hat{p}_0(x)}, -0.205, 0.241 \right\}. \tag{4.20}$$

It coincides with the nominal likelihood ratio on the interval $0.065 \leq x \leq 0.07$ and

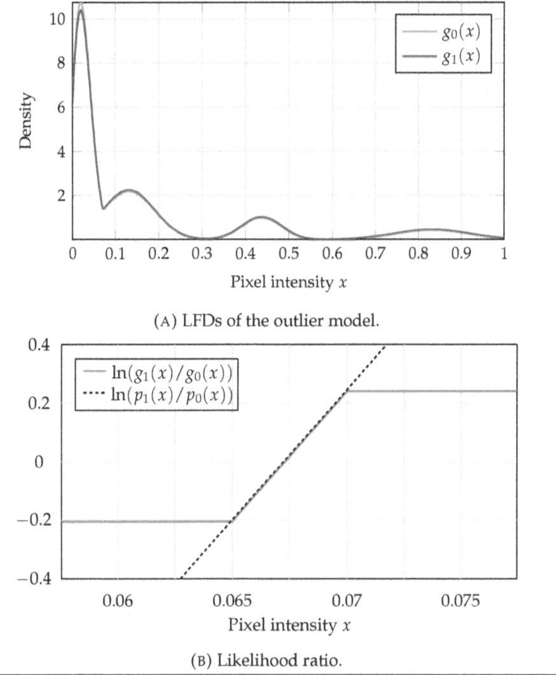

(A) LFDs of the outlier model.

(B) Likelihood ratio.

FIGURE 4.4: LFDs of the outlier model in (4.19) and their robust likelihood ratio in logarithmic scale.

is clipped at either -0.205 or 0.241 everywhere else. This clipping ensures that the influence of very small and very large x is limited and trusted only up to a certain level and illustrates a key feature of robust detectors, namely that a few outlying observations should not be able to outweigh the majority of the data.

4.5 Numerical Results

In this section, the performance of the robust detectors is evaluated and compared to that of non-robust approaches. Multiple FL-GPR images obtained from various viewpoints of the ROI, as described in Chapter 2, are used for the detection. More precisely, first, the LRT is performed on each radar image, then the resulting binary masks are multiplied pixel-by-pixel as in (4.14)–(4.16). That is, the final target regions are given by the intersections of the target regions of the individual images. For both the nominal and the robust tests the likelihood ratio thresholds γ in (3.6) and (4.14) are obtained by solving (3.7) for the estimated nominal and least favorable distributions, respectively. The detection results of the robust LRT detectors are evaluated for three different scenarios of environments, namely the low, mixed and

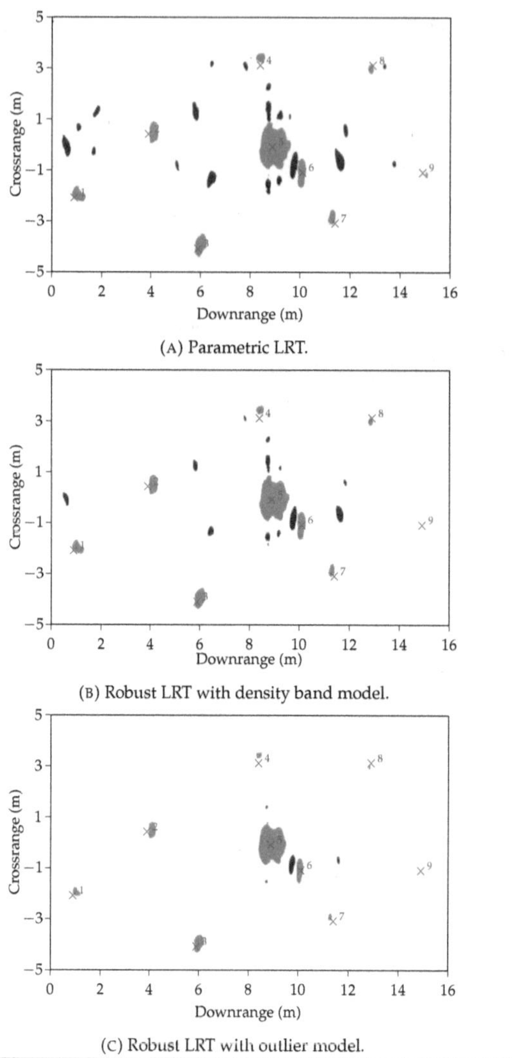

(A) Parametric LRT.

(B) Robust LRT with density band model.

(C) Robust LRT with outlier model.

FIGURE 4.5: Detection results in low roughness. Color coding: detected targets (red), FAs (black).

high surface roughness profiles. They are compared to the results obtained from the parametric LRT detector for the same FA rate constraint.

4.5.1 Low Surface Roughness

Fig. 4.5(a) shows the binary image resulting from the parametric LRT for a targeted FA rate of $\alpha = 0.05$. White pixels indicate a correct decision for H_0 (mines absent), red pixels indicate a correct decision for H_1 (mines present), and black pixels indicate an erroneous decision for H_1 (FA). The blue crosses indicate the central position of the true targets. As can be seen, all nine targets are successfully detected, but there exists several FAs as well. The large red area at 9 m downrange is due to the strong reflections from the metallic AT landmine positioned on the surface.

The result of the minimax LRT based on the density band model is depicted in Fig. 4.5(b). Here, the number of FAs is reduced significantly at the cost of one missed target, which is a buried plastic AT landmine located at the farthest downrange point. This result is to be expected: since the robust detector uses much weaker assumptions about the clutter distribution, it is less likely to confuse strong clutter echoes for targets. On the downside, this also makes the detector less sensitive towards weak target echoes.

Note that the detectors whose results are shown in Fig. 4.5(a) and Fig. 4.5(b) are both designed under the same FA constraint. Moreover, the effect of using LFDs is distinctly different from simply increasing the detection threshold of the parametric test. This can be seen from the fact that the target regions for all detected landmines are approximately of the same size. Increasing the threshold of the nominal test would also reduce the number of FAs, but at the cost of significantly smaller target regions. This effect can indeed be observed in Fig. 4.5(c), where the detection results of the minimax detector based on the second uncertainty model, the outlier model in (4.19), are depicted. The outlier model allows for significantly more uncertainty than the band model so that it is even less sensitive to clutter, while still correctly detecting all targets except for the plastic landmine 9. Hence, in this example, the outlier model can be argued to yield the best results. However, one should note that under the larger uncertainty model, the correctly detected targets are starting to be affected in the sense that the corresponding target regions are noticeably smaller and some landmines are "almost" missed. Given the grave consequences of having missed a target, this trade-off needs to be kept in mind when designing robust tests for landmine detection.

4.5.2 Mixed Surface Roughness

An advantage of minimax robust detectors is that they have a guaranteed performance under various conditions, without having to estimate the corresponding parameters. However, this also implies that robust detectors do not adapt to the observed data. Consequently, the advantages of robust detectors should be most pronounced in scenarios where adapting to the environment is difficult, i.e., the true

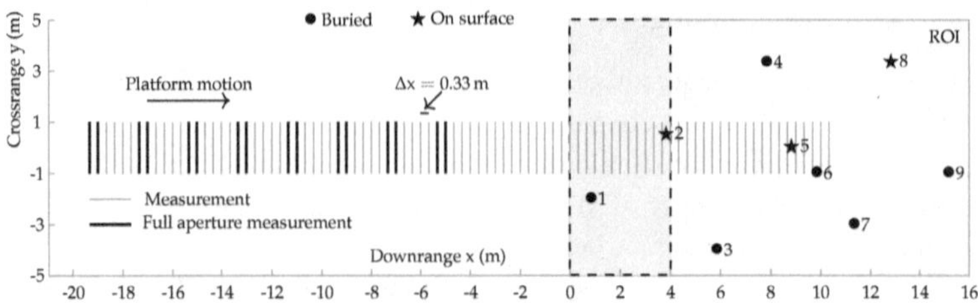

FIGURE 4.6: Top view of the FL-GPR measurement. Parallel lines indicate sensor array positions.

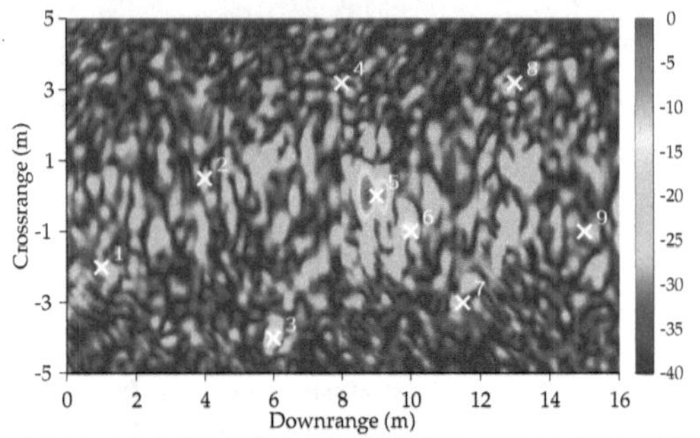

FIGURE 4.7: Tomographic image obtained from the farthest viewpoint of the ROI in high surface roughness environment. White crosses indicate targets.

distributions cannot be estimated reliably from the available training data. In FL-GPR, such situations occur when the statistical properties of the clutter are different for each measurement. This can happen, for example, in environments where the surface roughness of the ground changes frequently.

In order to simulate such a scenario, the ROI shown in Fig. 4.6 is divided into four areas whose roughness alternates between the low and the high profile. Each area has a size of $4\,\mathrm{m} \times 10\,\mathrm{m}$. Areas in the downrange from $4\,\mathrm{m}$ to $8\,\mathrm{m}$ and from $12\,\mathrm{m}$ to $16\,\mathrm{m}$ admit a high roughness profile ($h_{rms} = 1.6\,\mathrm{cm}$, $l_c = 14.93\,\mathrm{cm}$), while the remaining areas admit the low roughness profile ($h_{rms} = 0.8\,\mathrm{cm}$, $l_c = 14.26\,\mathrm{cm}$) that was used to generate the training data. The FL-GPR image for the higher roughness profile is shown in Fig. 4.7. The detection results for this scenario are shown in Fig. 4.8. All detectors generally exhibit an increased number of errors, especially on the segments of high roughness. However, it can be seen that the robust tests are significantly less affected by the changing clutter properties. In particular, the scenario based on the outlier uncertainty model manages to keep the number of FAs in check, without increasing the number of missed targets.

TABLE 4.2: Number of FAs and number of MDs for different surface
roughness.

LRT Detector	Low		Mixed		High	
	FA	MD	FA	MD	FA	MD
Parametric Model	22	0	40	1	41	3
Robust (Band Model)	12	1	19	1	17	3
Robust (Outlier Model)	4	1	8	1	12	3

4.5.3 High Surface Roughness

Finally, we consider a scenario where the entire investigation area admits a high
roughness profile. Fig. 4.9 shows the detection results for this scenario. Each detector commonly presents three missing targets, which are buried landmines number
4 and 9, and shell number 7. The increased number of errors is to be expected since
this is an effect of a mismatch between the true distribution of the measurement and
the assumed model. However, in comparison to the parametric model, the robust
detectors are preferable with fewer false alarms. Table 4.2 summarizes the performance of each detector in different rough surface environments. Accommodating
uncertainties in the distribution with the robust LRTs reduces the number of false
alarms significantly for all surface roughness levels. The robust LRT using the outlier model is preferable, with only one missed detection for both low and mixed
roughness profiles and the lowest number of false alarms among all detectors.

4.6 Summary

In this chapter, the problem of detecting landmines and unexploded ordnance in a
rough surface environment using FL-GPR has been investigated. The existence of
uncertainty in the image clutter and target distributions of the FL-GPR imagery is
examined by using bootstrap resampling and kernel density estimation methods. A
minimax robust LRT has been designed by first modeling the distributional uncertainties by density bands and then finding the corresponding least favorable densities. Two methods for constructing the uncertainty model as a neighborhood around
nominal distributions, i.e., the density band and the ϵ-contamination models, have
been investigated. The detection results of the robust detectors have been evaluated
for three different environments of low, mixed and high surface roughness profiles.
They have been compared to alternative parametric approaches for the same false
alarm rate constraint. Robust detectors have been shown to significantly reduce the
number of false alarms while maintaining a comparable detection rate. It has also
been shown that the uncertainty model needs to be chosen carefully in order to avoid
potential over-robustification.

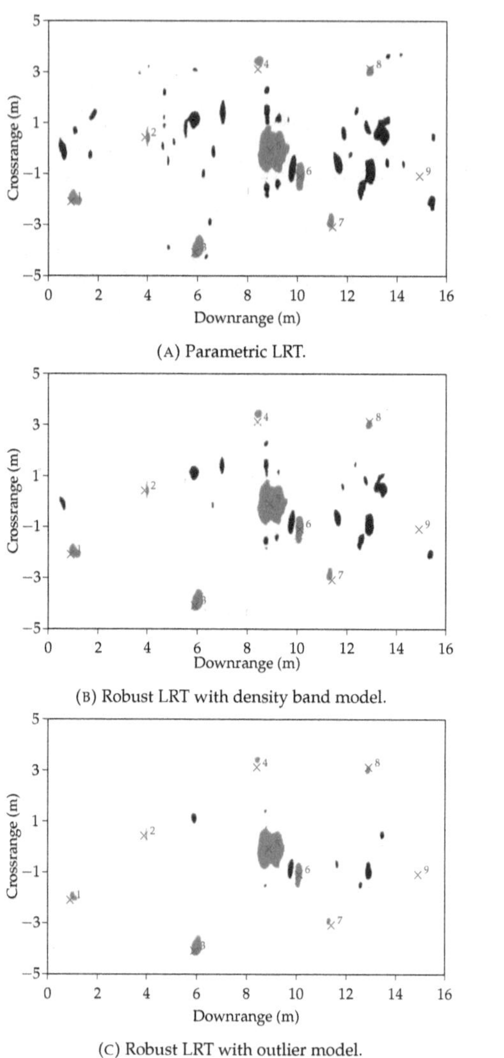

FIGURE 4.8: Detection results in mixed roughness. Color coding: detected targets (red), FAs (black).

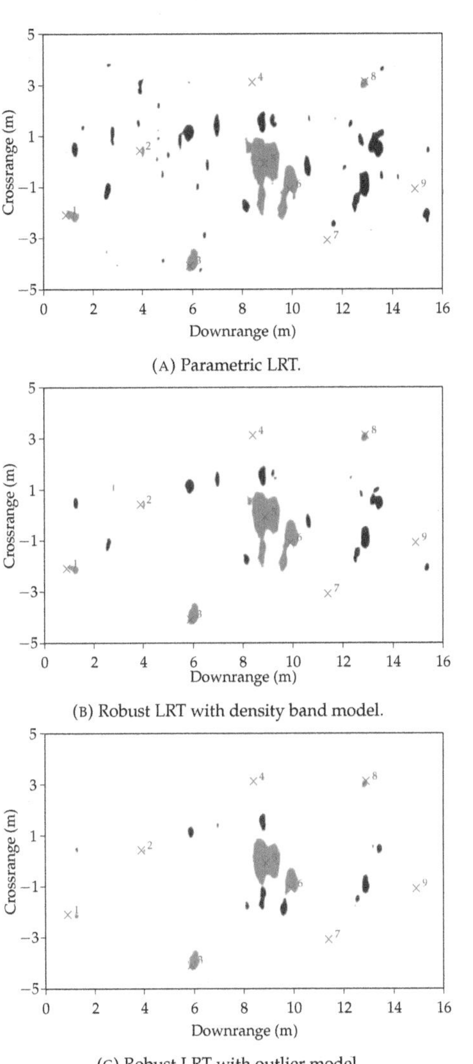

(A) Parametric LRT.

(B) Robust LRT with density band model.

(C) Robust LRT with outlier model.

FIGURE 4.9: Detection results in high roughness. Color coding: detected targets (red), FAs (black).

Chapter 5

Copula-Based Robust Approaches

In this chapter, copula robust approaches for landmine detection based on an LRT are considered. The detector is applied to FL-GPR images obtained from multiple viewpoints of the interrogation area. Different copula density functions are investigated in terms of their effectiveness to model the statistical dependence between multi-view images. The content presented in this chapter is partly taken from [PAZ20b; PAZ20a; PAZ21].

5.1 Motivation

The LRT approaches discussed in Chapters 3 and 4 assume the multi-view images to be statistically independent. In this chapter, this assumption is relaxed by developing a copula-based LRT detector which, similar to Chapter 4, is robust against deviations from the assumed statistical model. At the same time, it accounts for the statistical dependence between multiple images of the ROI from different viewpoints along the radar platform trajectory. More specifically, first, a density band within which the true pdf under each hypothesis is assumed to lie is constructed, and the test is designed to minimize the maximum error probability over all feasible distribution pairs within the two bands. Next, the design of the test is extended to capture the dependence structure between images from two consecutive viewpoints using a copula-based model. In order to perform a comprehensive analysis of the robust detector, three different copula density functions, namely, Gaussian, Clayton and Gumbel are investigated.

5.2 Robust Detection Approach

We now discuss a robust detection approach that builds on the statistical model presented in Section 4.2 for clutter and target distributions of FL-GPR images. M tomographic radar images, which are generated from the farthest to the closest viewpoints of the ROI, are considered. Under the null hypothesis H_0 (mines absent) and alternative H_1 (mines present), uncertainties in the distribution model are accounted for by the following hypothesis formulation:

$$H_s: \quad P_m \in \mathcal{P}_s, \quad \mathcal{P}_s = \left\{ p_s \mid \underline{p}_s(x) \leq p_s(x) \leq \bar{p}_s(x) \right\} \tag{5.1}$$

where $s \in \{0,1\}$, $m \in \{1,\ldots,M\}$ and P_m denotes the true distribution of pixel intensities x_m. The pixel distribution of an FL-GPR image under hypothesis H_s is assumed to be an element of an uncertainty set $\mathcal{P}_s$. The uncertainty set is modeled using Kassam's density band model [Kas81] such that the set of feasible distributions under hypothesis H_s is bounded by a lower density bound $\underline{p}_s$ and an upper density bound $\bar{p}_s$ as in (5.1). A robust LRT-based detector is designed by identifying the pair of densities that is least favorable in the sense that it minimizes the maximum error probability over all feasible density pairs inside the two bands, i.e., LFDs. A general form of the LFDs' functions for the band model is shown in (4.9).

In this chapter, a distributional uncertainty of the ϵ-contamination model is considered [Hub65]. That is, we assume a majority of the observations to be drawn from the parametric distribution model of Section 4.2, while an ϵ portion of the observations is drawn from a different probabilistic model representing outliers. Formally, the corresponding LFDs can be formed as

$$\begin{aligned} g_0(x) &= (1 - \epsilon_0)\, \hat{p}_0(x) + \epsilon_0\, h_0(x) \\ g_1(x) &= (1 - \epsilon_1)\, \hat{p}_1(x) + \epsilon_1\, h_1(x) \end{aligned} \tag{5.2}$$

where $\epsilon_s \in [0, 1/2)$ are the contamination ratios, $\hat{p}_s$ is the parametric density estimate given in (4.1) and h_s is the outlier density. The LFD g_s is lower bounded by a scaled-down parametric density estimate $\hat{p}_s$ and has no upper bound. This type of uncertainty can be expressed using the density band model by letting the upper bound $\bar{p}_s \to \infty$, so that the LFDs in (4.9) transform to

$$g_0(x) = \max\{\, a_0\, g_1(x),\ (1-\epsilon_0)\,\hat{p}_0(x)\,\}$$
$$g_1(x) = \max\{\, a_1\, g_0(x),\ (1-\epsilon_1)\,\hat{p}_1(x)\,\} \tag{5.3}$$

where the lower bound $\underline{p}_s$ is now of the form $(1-\epsilon_s)\,\hat{p}_s$, $s \in \{0,1\}$. Accordingly, the likelihood ratio of the LFDs can assume three possible values

$$\frac{g_1(x)}{g_0(x)} \in \left\{ \frac{1-\epsilon_1}{1-\epsilon_0}\frac{\hat{p}_1(x)}{\hat{p}_0(x)}, a_1, \frac{1}{a_0} \right\}. \tag{5.4}$$

The first value corresponds to the case where the LFDs coincide with the lower density bounds, $g_s = \underline{p}_s$. For an identical contamination under both hypotheses, i.e., $\epsilon_0 = \epsilon_1$, this value coincides with the likelihood ratio of the parametric densities $\hat{p}_1(x)/\hat{p}_0(x)$. The two remaining values in (5.4) imply that $g_s > \underline{p}_s$. This test statistic leads to a *clipping* of the parametric test statistic $(\hat{p}_1(x)/\hat{p}_0(x))$, thereby ensuring that the influence of x is bounded and exerted only up to a certain level. In this manner, a key feature of robust detectors is illustrated, i.e., they do not outweigh the majority of the data by a few outlying observations. Therefore, the robust test is a pixel-wise LRT of the form (3.6), but with the conditional densities (p_0, p_1) replaced by a pair of LFDs (g_0, g_1).

5.3 Copula-Based Multi-View Fusion

The robust test described in Section 5.2 is optimized by taking into account statistical dependence between the multi-view images. A copula model is employed to incorporate such dependencies into the robust test.

5.3.1 Test Statistic of Consecutive Images

The copula-based model is supported by *Sklar's Theorem* [Sch91]. Let (x_m, x_{m+1}) be pixel intensities of two radar images corresponding to two consecutive viewpoints. The joint distribution of the pixel intensity of these radar images $P_s(x_m, x_{m+1})$ under hypothesis H_s is defined by joining its marginal distributions $P_{1,s}(x_m)$, $P_{2,s}(x_{m+1})$ and a unique copula function $C_s(\cdot)$ as follows:

$$P_s(x_m, x_{m+1}) = C_s(P_{1,s}(x_m), P_{2,s}(x_{m+1})), \quad s \in \{0,1\}. \tag{5.5}$$

The copula C_s is a bivariate distribution function with uniform marginals. That is, $U_{1,s} = P_{1,s}(x_m) \sim \mathcal{U}(0,1)$ and $U_{2,s} = P_{2,s}(x_{m+1}) \sim \mathcal{U}(0,1)$ by probability transformation, as detailed in Appendix B. Conversely, given $(U_{1,s}, U_{2,s}) \sim C_s$ and univariate distributions $P_{1,s}, P_{2,s}$, we have

$$
\begin{aligned}
C_s(u_{1,s}, u_{2,s}) &= C_s(P_{1,s}(P_{1,s}^{-1}(u_{1,s})), P_{2,s}(P_{2,s}^{-1}(u_{2,s}))) \\
&= C_s(P_{1,s}(x_m), P_{2,s}(x_{m+1}))
\end{aligned}
\tag{5.6}
$$

which defines the joint distribution with marginals $P_{r,s}, r = 1, 2$, where $u_{r,s} \in [0,1]$. The joint density is obtained by taking the second-order derivative of (5.5) as

$$
\begin{aligned}
p_s(x_m, x_{m+1}) &= \frac{\partial^2}{\partial x_m \partial x_{m+1}} C_s(P_{1,s}(x_m), P_{2,s}(x_{m+1})) \\
&= c_s(P_{1,s}(x_m), P_{2,s}(x_{m+1})) \, p_{1,s}(x_m) \, p_{2,s}(x_{m+1})
\end{aligned}
\tag{5.7}
$$

which is the product of its marginal densities $p_{1,s}$ and $p_{2,s}$, *weighted* by a copula density $c_s(\cdot)$ to account for dependence between radar images. Using this model, we can examine the characteristics of the underlying marginal distributions separately from the dependence structure between radar images specified by a suitable copula.

By substituting the LFDs for the marginal densities in (5.7), a robust copula-based test statistic of two consecutive radar images is developed as

$$
\begin{aligned}
L_{CR} = \ln \frac{g_1(x_m)}{g_0(x_m)} &+ \ln \frac{g_1(x_{m+1})}{g_0(x_{m+1})} \\
&+ \ln \frac{c_1(P_{1,1}(x_m), P_{2,1}(x_{m+1}))}{c_0(P_{1,0}(x_m), P_{2,0}(x_{m+1}))}.
\end{aligned}
\tag{5.8}
$$

The test statistic L_{CR} has two main properties. First it is robust against statistical model deviations of radar images. Second, it optimizes performance by taking into account any statistical dependence that may be present between radar images. Here, the dependence structure is constructed by selecting three prominent parametric copulas, namely Gaussian, Gumbel and Clayton, which represent dependence models of the symmetric, upper tail and lower tail behaviors, respectively [IVD11; Bou+09; Nel99]. The definitions of the aforementioned copula functions are provided in Table 5.1, where Φ and Φ_{ρ_s} denote the respective standard univariate and bivariate Gaussian distributions with correlation coefficient ρ_s, ς_s describes the parameter of the Gumbel copula and τ_s is the parameter of the Clayton copula function. The product copula represents the case of zero dependence between the two images. This type of copula is implicitly chosen when assuming independence between images from different views, as we did in Chapters 3 and 4. The simulation results indicate, that the incorporation of a more dedicated copula function increases the performance.

TABLE 5.1: Copula functions.

Model	Function $C_s(u_{1,s}, u_{2,s})$
Gaussian C_s^{Ga}	$\Phi_{\rho_s}(\Phi^{-1}(u_{1,s}), \Phi^{-1}(u_{2,s}))$
Gumbel C_s^{Gu}	$e^{-\left((-\ln u_{1,s})^{\varsigma_s} + (-\ln u_{2,s})^{\varsigma_s}\right)^{\frac{1}{\varsigma_s}}}$
Clayton C_s^{Cl}	$(u_{1,s}^{-\tau_s} + u_{2,s}^{-\tau_s} - 1)^{-\frac{1}{\tau_s}}$
Product C_s^{Pro}	$u_{1,s}u_{2,s}$

5.3.2 Copula Parameter Estimation

The selected copula functions have one parameter each that governs the level of dependency between two successive radar images. In order to estimate the parameter of the copula function, a maximum likelihood-based approach known as the inference function for margins (IFM) is used [JX96]. Let $\theta_{C,s}$ denote the parameter controlling the dependency of any copula function and $\{\theta_{r,s}, r = 1,2\}$ be the set of parameters of the marginal densities under hypothesis H_s. From (5.7), it follows that the corresponding log-likelihood based on N samples can be written as

$$\sum_{n=1}^{N} \ell_{C,s}(\theta_{C,s}; P_{1,s}(x_{m,n}; \theta_{1,s}), P_{2,s}(x_{m+1,n}; \theta_{2,s}))$$

$$+ \sum_{n=1}^{N} (\ell_{1,s}(\theta_{1,s}; x_{m,n}) + \ell_{2,s}(\theta_{2,s}; x_{m+1,n})) \tag{5.9}$$

where

$$\ell_{C,s}(\theta_{C,s}; u_{1,s}, u_{2,s}) = \ln c_s(u_{1,s}, u_{2,s}; \theta_{C,s})$$
$$\ell_{r,s}(\theta_{r,s}; x) = \ln p_{r,s}(x; \theta_{r,s}), \quad r = 1,2. \tag{5.10}$$

With the IFM method, $\theta_{C,s}$ and $\{\theta_{r,s}, r = 1,2\}$ can be estimated by a two-step procedure, which makes the computation less expensive than the conventional maximum likelihood approach; the latter requires the joint estimation of marginal and copula parameters. The parameters of marginal distributions are first estimated from the second sum in (5.9), and then the copula parameter is estimated from the first sum. This approach is also known as maximum pseudo-likelihood estimation (MPLE) [Hau16; GGR95].

Since the distributions under null hypothesis and alternative are only known approximately in robust testing, the marginal distributions are replaced with their sample estimates, given by

$$\hat{P}_{1,s}(x_m) = \frac{1}{N} \sum_{n=1}^{N} I(x_{m,n} \leq x_m), \quad \hat{P}_{2,s}(x_{m+1}) = \frac{1}{N} \sum_{n=1}^{N} I(x_{m+1,n} \leq x_{m+1}) \tag{5.11}$$

where $I(A)$ is the indicator function of event A. Thus, from the first term in (5.9), the MLEs of the copula parameters are obtained as

$$\hat{\rho}_s = \arg\max \sum_{n=1}^{N} \ln c_s^{\mathrm{Ga}}(\hat{P}_{1,s}(x_{m,n}), \hat{P}_{2,s}(x_{m+1,n}); \rho_s) \tag{5.12}$$

$$\hat{\varsigma}_s = \arg\max \sum_{n=1}^{N} \ln c_s^{\mathrm{Gu}}(\hat{P}_{1,s}(x_{m,n}), \hat{P}_{2,s}(x_{m+1,n}); \varsigma_s) \tag{5.13}$$

$$\hat{\tau}_s = \arg\max \sum_{n=1}^{N} \ln c_s^{\mathrm{Cl}}(\hat{P}_{1,s}(x_{m,n}), \hat{P}_{2,s}(x_{m+1,n}); \tau_s). \tag{5.14}$$

For more details on the copula density functions c_s^{Ga}, c_s^{Gu} and c_s^{Cl} and the MLE, see Appendix C.

5.3.3 Multi-View Fusion

Decision-making with a set of M consecutive FL-GPR images proceeds as follows. First, a local decision is assigned to each pair of radar images corresponding to the m-th and $(m+1)$-th viewpoints of the ROI according to the following rule:

$$d_m(i,j) = \begin{cases} 1, & L_{\mathrm{CR}}(x_m(i,j), x_{m+1}(i,j)) > \gamma \\ 0, & L_{\mathrm{CR}}(x_m(i,j), x_{m+1}(i,j)) \le \gamma \end{cases} \tag{5.15}$$

where

$$L_{\mathrm{CR}}(x_m(i,j), x_{m+1}(i,j)) = \ln \frac{g_1(x_m(i,j))}{g_0(x_m(i,j))} + \ln \frac{g_1(x_{m+1}(i,j))}{g_0(x_{m+1}(i,j))}$$
$$+ \ln \frac{\hat{c}_1(\hat{P}_{1,1}(x_m(i,j)), \hat{P}_{2,1}(x_{m+1}(i,j)))}{\hat{c}_0(\hat{P}_{1,0}(x_m(i,j)), \hat{P}_{2,0}(x_{m+1}(i,j)))}, \tag{5.16}$$

the copula density $\hat{c}_s$ is an estimate of the selected copula function under hypothesis H_s, the threshold γ is determined using (3.7) based on a nominal FA rate α, and $m = 1, \ldots, M-1$.

Next, a multi-view hard fusion is applied to the M radar images as

$$D(i,j) = \prod_{m=1}^{M-1} d_m(i,j) \quad i = 1, \ldots, N_x,\ j = 1, \ldots, N_y \tag{5.17}$$

Hence, a single binary image $D(i,j)$ is obtained through this multi-view fusion procedure. The alternative H_1, i.e., presence of a mine, is declared at all pixels with significant test statistics L_{CR} for all $M-1$ image pairs from the farthest to the closest viewpoints of the ROI.

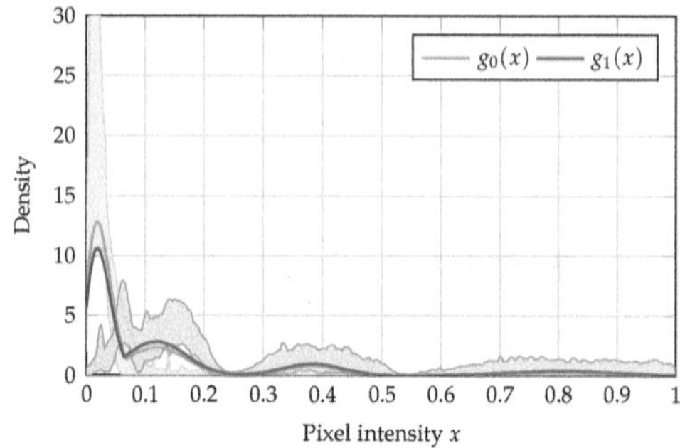

FIGURE 5.1: LFDs (g_0, g_1) from the model given in (5.3) with the confidence bands.

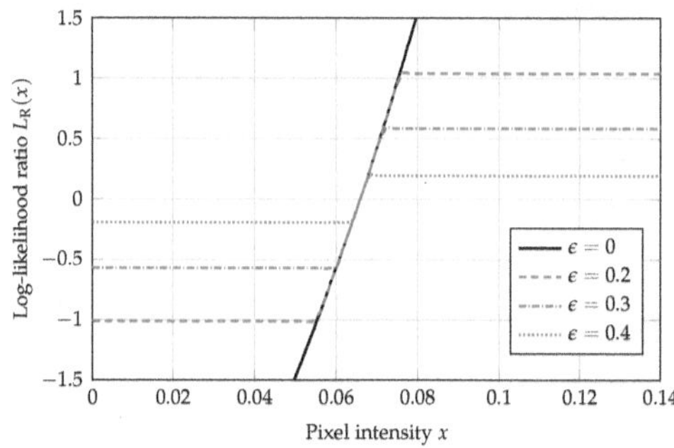

FIGURE 5.2: Log-likelihood ratio of the parametric densities ($\epsilon = 0$) and the LFDs for different contamination ratios $\epsilon > 0$.

5.4 Numerical Results

The copula-based robust detector is applied to FL-GPR imagery for the detection of landmines buried at shallow depths in the ground with varying surface roughness profiles. The performance of the detection techniques for Gaussian, Gumbel and Clayton copulas is evaluated in terms of the pixel-level test accuracy. For comparison, the accuracies for the parametric and robust counterparts, designed under the assumption of statistically independent multi-view images, are provided.

5.4.1 Image Test Statistics

In order to estimate the LFDs (g_0, g_1), two sets of clutter and target samples are used. These sets are obtained from the target-free and clutter-free FL-GPR images as discussed in Section 3.4.2. Fig. 5.1 shows the LFDs (g_0, g_1) from the model given in (5.3) for the contamination ratio of $\epsilon = 0.4$ with the confidence density bands from Section 4.2. This value was also examined for the outlier ϵ-contamination model in Chapter 4. Even though the assumption of only 60% pixels following the parametric model is rather pessimistic, it complies with the idea of a minimax robust approach, which prepares for a worst-case condition. Clearly, the LFDs are difficult to distinguish since they are similarly imposed by construction. The corresponding log-likelihood ratio of the LFDs for $\epsilon = 0.4$ is illustrated in Fig. 5.2. It coincides with the likelihood ratio using the parametric densities ($\epsilon = 0$) for $0.065 \leq x \leq 0.07$ and is clipped at either -0.205 or 0.241 outside this interval. As shown in Fig. 5.2, the lower the contamination ratio, the wider is the non-clipped interval, which implies that more pixels are assumed to follow the parametric distribution model. This clipping of the test statistic guarantees that a small fraction of outliers will not override the statistical evidence of the majority.

The dependence parameters of the selected copula functions are estimated using 3206×2 mine samples and $830\,592 \times 2$ clutter samples obtained from the clutter-free and target-free images, corresponding to the two closest consecutive viewpoints. Using the MPLE method, the parameters of the Gaussian, Gumbel and Clayton copulas are estimated as $(\hat{\rho}_0, \hat{\rho}_1) \approx (0.897, 0.996)$, $(\hat{\varsigma}_0, \hat{\varsigma}_1) \approx (3.704, 11.024)$ and $(\hat{\tau}_0, \hat{\tau}_1) \approx (2.124, 15.874)$, respectively. As dictated by the estimated values of the copula parameters, the dependence under hypothesis H_1 (mines present) is higher than that under the null hypothesis H_0 (mines absent) for all three copula models. These statistical characteristics are incorporated into the LRT as in (5.15)–(5.17) to enhance detection performance.

5.4.2 Detection Results

First, the robust detector is applied to a mixed surface roughness scenario. The ROI of Fig. 2.5 is divided into four segments with surface roughness alternating between the low and the high profiles. Each segment has dimensions of $4\,\text{m} \times 10\,\text{m}$. Areas along downrange from $4\,\text{m}$ to $8\,\text{m}$ and from $12\,\text{m}$ to $16\,\text{m}$ have a high roughness profile, while the remaining areas correspond to the low roughness profile. Corresponding detection results based on parametric and robust tests for an FA rate of 5% are presented as binary images in Figs. 5.3–5.5. The blue crosses mark the true central positions of the targets. White and red pixels indicate respective correct decisions for H_0 and H_1. The black pixels indicate an erroneous decision for H_1 (FA). For multi-view fusion, a total of $M = 11$ radar images from the farthest to the closest viewpoints of the ROI is used.

Fig. 5.3 shows the detection results assuming statistical independence across multiple images. The binary image resulting from the parametric LRT is shown in Fig. 5.3(a). As we can see, all mines are detected with the exception of a buried plastic AT landmine (target 9) located at the farthest downrange in the ROI. The large red area at around $9\,\text{m}$ downrange corresponds to the strong return from the metallic AT landmine placed on the surface (target 5). A high number of FAs are present, which can be attributed to not only the sensitivity of the parametric test to

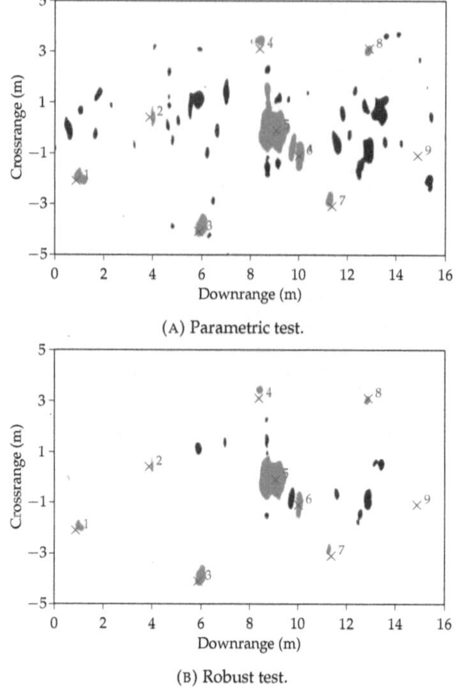

(A) Parametric test.

(B) Robust test.

FIGURE 5.3: Detection results obtained under the assumption of independence across images. Color coding: detected targets (red), FAs (black).

deviations in the assumed distributions, but also the invalidity of the statistical independence assumption across multiple views. Fig. 5.3(b) shows the binary image from the robust test based multi-view fusion [Pam+20b]. It uses the robust test statistic shown in Fig. 5.2 with a contamination ratio of $\epsilon = 0.4$. We see that the number of FAs is reduced significantly (by 73% approximately) as compared to those from the parametric test.

Fig 5.4 presents the detection results from the robust copula-based tests with a contamination ratio of $\epsilon = 0.4$. As shown in Fig. 5.4(a), the robust copula-based test with the Gaussian copula function results in a reduced number of FAs as compared to those as in Fig. 5.3(b) from the basic robust test under the independence assumption. Further, it reduces the sidelobe contributions of target 5. The result corresponding to the Gumbel copula function appears to be identical to Fig. 5.3(b). The lack of improvement in this case can be attributed to the fact that the Gumbel copula puts a lot of weight on the upper tail dependence. The pixel distributions, on the other hand, are heavier for lower than on the higher values under both hypotheses, as seen in Fig. 5.1. Finally, a significant reduction in the number of FAs is achieved by

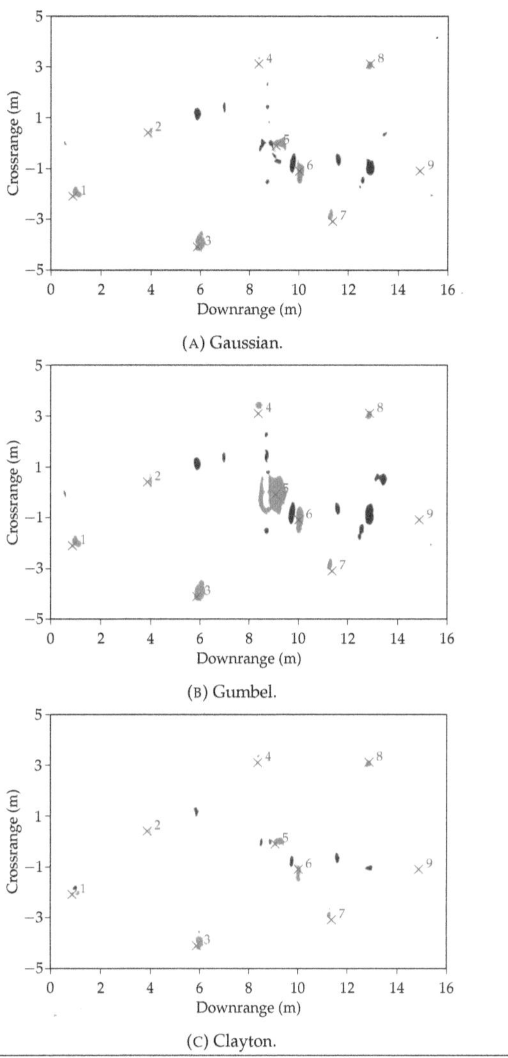

FIGURE 5.4: Detection results of robust copula-based tests for different copulas. Color coding: detected targets (red), FAs (black).

using the Clayton copula function, as shown in Fig. 5.4(c). Also, Fig. 5.5 depicts the binary images resulting from the robust copula-based test using the Clayton copula function but for lower contamination ratios as compared to that in Fig. 5.4(c). Note that as the contamination ratio decreases, the likelihood ratio of the robust test resembling the parametric test increases, as seen in Fig. 5.2; the latter causes a higher number of FAs. This is also validated by comparing the binary images in

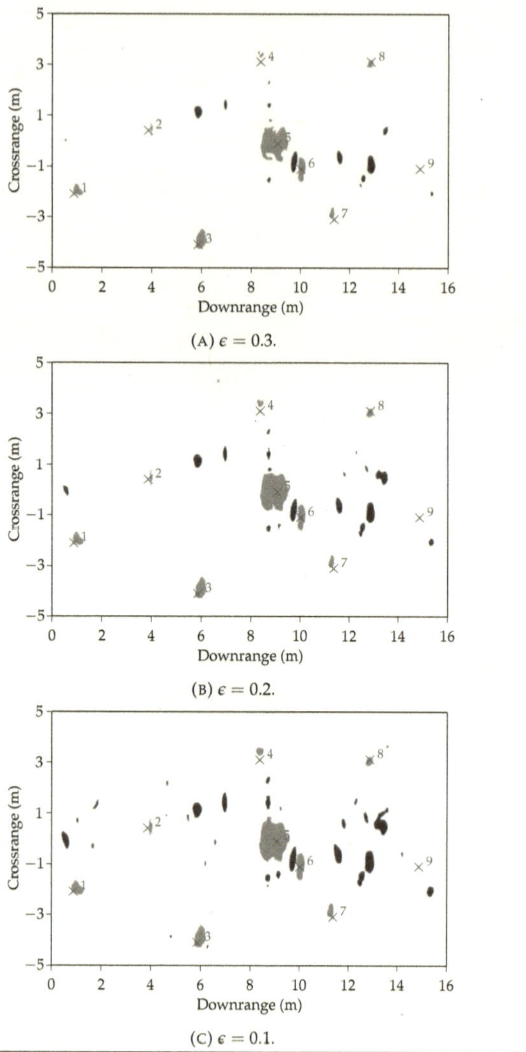

FIGURE 5.5: Detection results of robust copula-based tests from different contamination ratios. Color coding: detected targets (red), FAs (black).

Figs. 5.4(c), 5.5(a), 5.5(b), 5.5(c) for $\epsilon = 0.4, 0.3, 0.2, 0.1$, respectively.

The performance of the robust copula-based LRT is tabulated in terms of the FA reduction rate $v(\%)$ with respect to the number of FAs from the parametric test for different rough surface environments and contamination ratios. Table 5.2 deals with

TABLE 5.2: FA reduction rate $v(\%)$ for mixed low (L) and high (H) surface roughness and for different contamination ratios.

	LHLH			HLHL		
ϵ	0.2	0.3	0.4	0.2	0.3	0.4
Gaussian	52	68	78	51	67	76
Gumbel	50	66	76	46	64	71
Clayton	64	78	91	52	71	83
Product	35	64	73	37	50	65

TABLE 5.3: FA reduction rate $v(\%)$ for low or high surface roughness and for different contamination ratios.

	Low			High		
ϵ	0.2	0.3	0.4	0.2	0.3	0.4
Gaussian	40	63	76	44	65	76
Gumbel	37	60	71	43	63	75
Clayton	50	68	80	54	68	81
Product	33	57	69	37	60	72

the mixed roughness cases, namely, LHLH and HLHL. The latter indicates an alternating surface roughness where the ROI along downrange from 0 m to 4 m and between 8 m to 12 m has a high roughness profile, while the remaining areas correspond to the low roughness profile. On the other hand, Table 5.3 summarizes the results corresponding to the individual cases where the ROI has either a low or high surface roughness. The independence assumption condition is covered by the product copula in Tables 5.2 and 5.3. It is evident that incorporation of the statistical dependence between images into the robust test increases the FA reduction rate for all surface roughness profiles. Further, the performance of the robust copula-based test depends on the contamination ratio in the distribution. The robust copula-based test using the Clayton copula function with a contamination ratio of $\epsilon = 0.4$ is preferable with the highest FA reduction rate amongst all detectors.

5.4.3 Performance Evaluation

The performance of the robust copula-based test is further evaluated over a wider range of probability of FAs. To provide a fair comparison, dependence structures are also incorporated into the parametric LRT. The test accuracy $\kappa(\epsilon)$ is used as a measure of the test performance. The accuracy is defined as the proportion of correct

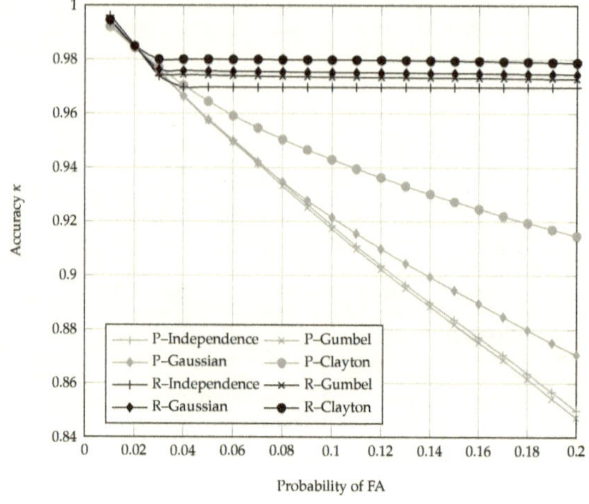

FIGURE 5.6: Performance gain of the parametric (P) and the robust (R) tests when using the copula dependence model.

decisions, i.e., correctly classified pixels for each hypothesis, among the total image pixels ($N_x \times N_y$). Fig. 5.6 shows the performance comparison of the parametric test and the robust test for a contamination ratio of $\epsilon = 0.4$. Compared to the test that assumes independent multi-view images, incorporating dependence structures with a well-suitable copula increases the accuracy of both the parametric and the robust tests. The highest accuracy is achieved by using the Clayton copula function for both tests. More importantly, compared to the parametric test that comes without statistical performance guarantees, the robust test provides a guaranteed accuracy exceeding 96% over the entire range of the required FA rates, as depicted in Fig. 5.6.

Next, the performance of the proposed robust detectors vs. $\epsilon \in [0, 1/2)$, for a FA rate of 5%, is evaluated. The empirical probability of detection $P_D(\epsilon)$ is measured by counting the number of pixels at which the alternative hypothesis H_1 is correctly accepted. The characteristic performance of the robust copula-based detectors is summarized in Fig. 5.7, by plotting the empirical P_D versus the accuracy κ for the considered contamination ratios. Each point on the plot corresponds to a value of (P_D, κ) for an assumed contamination ratio ϵ. Note, that P_D does not reach unity due to one missed detection of the buried plastic AT landmine located at the farthest downrange point. Further, as ϵ decreases, P_D increases, but the accuracy κ decreases. The performance characteristics shown in Fig. 5.7 can be used to obtain a suitable contamination ratio for the robust copula-based detectors via

$$\underset{\epsilon \in [0, 1/2)}{\arg \max} \ (1 - w)\kappa(\epsilon) + wP_D(\epsilon) \tag{5.18}$$

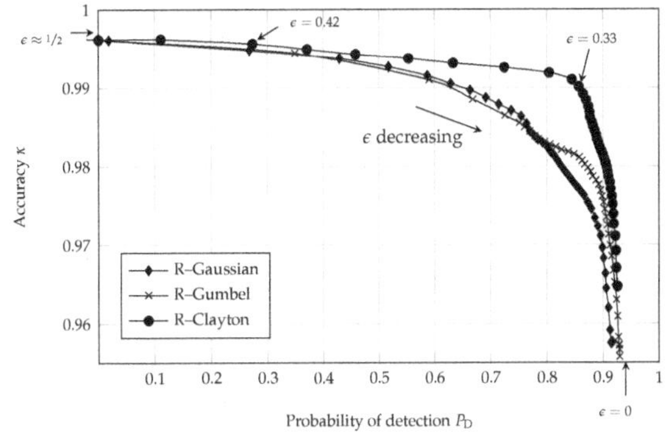

FIGURE 5.7: Robust copula-based detector characteristics.

where $w \in (0,1)$ can be chosen accordingly to weight each performance measure. For $w = 0.2$, we obtain suitable contamination ratios of $\epsilon = (0.17, 0.21, 0.33)$ for the Gaussian, Gumbel and Clayton copulas, respectively. A value of $\epsilon \approx 0.42$ is observed as the maximum contamination ratio for detecting eight mines with significantly high FA reduction.

Appendix D provides additional evaluation of the robust detector performance over different amounts of uncertainty in the distributions, which is realized by shifting the lower density bound $\underline{p}_s$ under hypothesis H_s, $s = \{0, 1\}$.

5.5 Summary

In this chapter, copula-based robust approaches have been presented for enhanced detection of landmines whose signatures in the FL-GPR image domain are obscured by clutter generated from rough ground surface environments. A robust detection technique based on the LRT, is applied to multi-view radar images of the ROI. The robust test was designed to account for the distribution uncertainty under each hypothesis and by incorporating statistical dependencies across the images via copula functions. The detector performance was validated for varying ground surface roughness profiles. Superiority of the robust copula-based detector over alternative LRT techniques was demonstrated by means of a higher test accuracy and a significant reduction in the FA rate. The robust detector was shown to provide a guaranteed performance over a wide range of FA rates.

Chapter 6

Conclusions and Future Directions

In this book, the problem of detecting landmines using FL-GPR in rough surface environments has been considered. Detection techniques have been developed and applied to multi-view FL-GPR images. Uncertainty models have been constructed to capture feasible targets and clutter statistics in the radar image. The detection techniques have been designed to perform well under uncertainty of the radar image statistics. The design of the detectors have been extended to capture statistical dependence between two consecutive images using copula-based models. The book has set a focus on robust techniques and optimized detection performance.

A summary and the main conclusions of this book work are provided in Section 6.1. Finally, Section. 6.2 provides an outlook for possible future work.

6.1 Conclusions

6.1.1 Non-Robust Detection Technique

Detectors based on a simple threshold scheme and a likelihood-ratio test have been presented. The detectors have been applied to tomographic FL-GPR images obtained from multiple viewpoints of the region of interest. A parametric family of distribution has been employed for the likelihood-ratio test detector, providing nominal statistics of the radar image. Multi-view schemes based on the likelihood-ratio test have been used to obtain a desired performance of a reduced number of false alarms and an increased detection power. The superiority of the multi-view approaches over the single-view imaging case has been demonstrated through its ability to reduce the influence of clutter on the detection performance.

6.1.2 Robust Detection Technique

Uncertainties in the image clutter and target distributions of the FL-GPR imagery have been observed by using bootstrap resampling and kernel density estimation methods. A minimax robust likelihood-ratio test has been designed by first modeling the distributional uncertainties by density bands and then finding the corresponding least favorable densities. Two methods for constructing the uncertainty model as a neighborhood around nominal distributions, i.e., the density band and the ϵ-contamination models, have been investigated. The detection results of the robust detectors have been evaluated for three different surface environments of low, mixed and high roughness profiles. They have been compared to alternative parametric approaches for the same nominal false alarm rate. Robust detectors have been shown to significantly reduce the number of false alarms while maintaining a comparable detection rate. In order to avoid potential over-robustification, it has been shown that the uncertainty model needs to be chosen carefully .

6.1.3 Copula-Based Robust Approach

Statistical dependencies between multi-view images of the interrogation area have been examined. Copula-based robust approaches have been presented for enhanced detection of landmines whose signatures in the FL-GPR image are obscured by clutter. A robust detection technique, which is based on the likelihood-ratio test, has been applied to multi-view radar images of the region of interest. The robust test has been designed to account for the distribution uncertainty under each hypothesis and by incorporating statistical dependencies across the images via copula-based models. Different copula functions have been investigated in terms of their effectiveness in incorporating the dependence structure between multi-view images into the test. The detector performance has been validated for varying ground surface roughness profiles. Superiority of the copula-based robust detector over alternative likelihood-ratio test detectors has been demonstrated through a higher test accuracy and a significant reduction in the false-alarm rate. The robust detector has been shown to posses a guaranteed test performance over a wide range of nominal false alarm rates.

6.2   Future Directions

Up to now, the humanitarian demining process has depended highly on manual deminers equipped with metal detectors. The manual technique is still conducted, at the expense of high risk operation. This book work has significantly advanced the performance of FL-GPR, which can potentially contribute towards modernizing efforts in humanitarian demining action. However, further research in this area is needed.

6.2.1   Distributional Uncertainty

In the construction of uncertainty sets of targets and clutter in the FL-GPR image, we have made assumptions to simplify the problem. The assumptions, such as the ground being homogeneous and a target having single material composition, need to be relaxed for practical application in landmine detection. A realistic assumption of various ground and target material compositions should be considered. This realistic assumption can provide a sufficiently reliable estimate of the uncertainty sets. In this work, the use of uncertainty classes based on ϵ-contamination and density band models was effective; however, there exists another class of uncertainty based on distance measures such as K–L divergence, alpha divergence and other types of f-divergences.

6.2.2 Dependence Modeling

Incorporating statistical dependence between radar images using a copula-based model has been shown to yield increased detection performance. In this book, the dependence structure between images from two consecutive viewpoints has been successfully captured by using Gaussian and Archimedean bivariate copula families; however, other families of multivariate copula exist, such as C–vine, D–vine and R–vine copulas, which are more flexible for modeling high-dimensional dependence structures [Zha+19]. Investigating the dependence structure of more than two consecutive viewpoints of the radar image using the family of vine copulas is a potential area of further research to improve the effectiveness of the technique. In addition, the question of how to properly model the dependence between random variables under model uncertainty warrants further theoretical research.

6.2.3 Multiple Testing Correction

The detection approach can be extended by considering the problem of multiple comparisons, where a set of statistical tests is considered continuously. Instead of controlling the probability of false alarms, an alternative performance measure such as false discovery rate can be used. A brief discussion of the measure has been provided in Section 3.3.2, where a simple controlling error procedure was examined for a single-view FL-GPR image. Following this technique can potentially yield a more meaningful measure of detection error in the FL-GPR image. As a first step in this direction, an approach was proposed in [Pam+20a].

Appendices

A   Measurement Configuration

This appendix provides top views of the FL-GPR measurement configuration, indicating sixteen platform positions integrated to generate the image segments (dashed line rectangle) corresponding to different viewpoints.

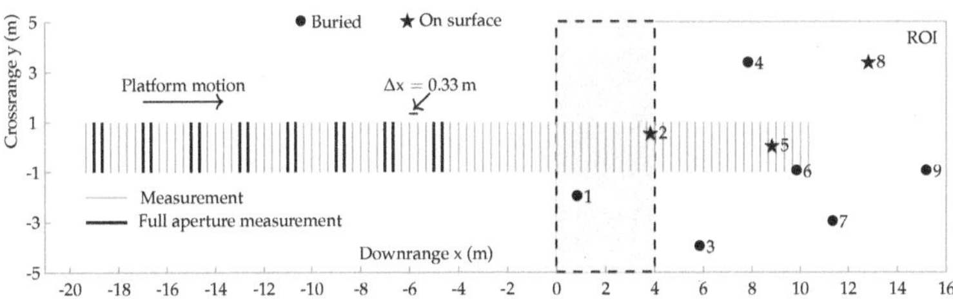

FIGURE 1: Measurement configuration of the first image segment corresponding to the second viewpoint.

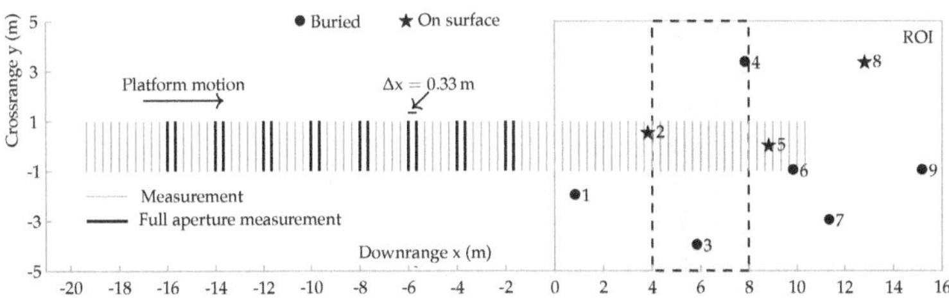

FIGURE 2: Measurement configuration of the second image segment corresponding to the first viewpoint.

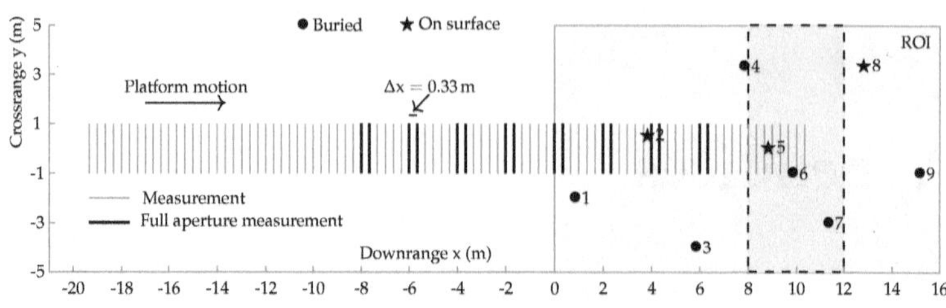

FIGURE 3: Measurement configuration of the third image segment corresponding to the eleventh viewpoint.

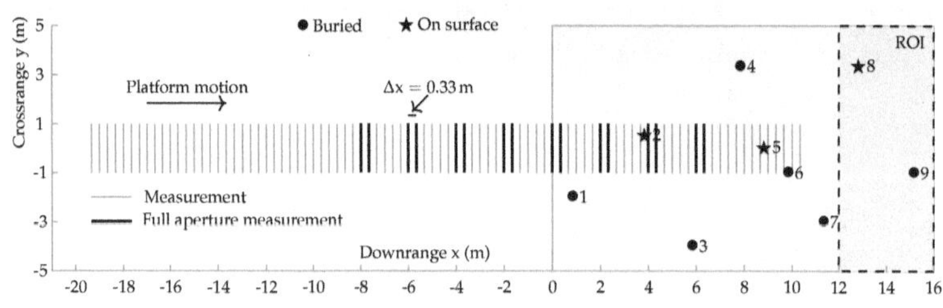

FIGURE 4: Measurement configuration of the fourth image segment corresponding to the first viewpoint.

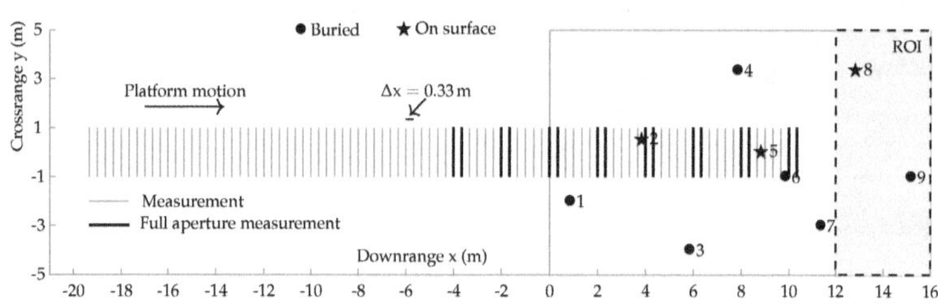

FIGURE 5: Measurement configuration of the fourth image segment corresponding to the eleventh viewpoint.

B Key of Copula Applications

This appendix provides keys to the copula-based dependence models, namely probability and quantile transformations which are described as follows.

- *Probability transformation*: Let $X \sim P_s$ and P_s be a continuous cumulative distribution function (cdf) of the FL-GPR image under hypothesis H_s. Then

$$
\begin{aligned}
P\left(P_s(X) \leq u\right) &= P\left(P_s^{-1}(P_s(X)) \leq P_s^{-1}(u)\right) \\
&= P\left(X \leq P_s^{-1}(u)\right) \\
&= P_s\left(P_s^{-1}(u)\right) = u, \ u \in [0,1]
\end{aligned}
\tag{B1}
$$

which shows that $P_s(X) \sim \mathcal{U}(0,1)$.

- *Quantile transformation*: Let $U \sim \mathcal{U}(0,1)$ and P_s any cdf of an FL-GPR image under hypothesis H_s. Then

$$
\begin{aligned}
P\left(P_s^{-1}(U) \leq x\right) &= P\left(P_s(P_s^{-1}(U)) \leq P_s(x)\right) \\
&= P\left(U \leq P_s(x)\right) \\
&= P_s(x), \ x \in \mathbb{R}
\end{aligned}
\tag{B2}
$$

which shows that $X = P_s^{-1}(U) \sim P_s$.

Both transformations are the basic principles of the copula model which allow us to go from $x \in \mathbb{R}^2$ to $u \in [0,1]^2$ and back.

C   Copula Parameter Estimation

A maximum likelihood-based approach known as the method of inference functions for margins IFM [JX96] is used to estimate dependence parameters of the Gaussian, Gumbel and Clayton copulas. First, by taking the second order derivative of the selected copula functions as shown in Table 5.1, the corresponding copula densities can be obtained as

$$c_s^{\mathrm{Ga}}(u_{1,s}, u_{2,s}) \tag{C1}$$

$$= \frac{\partial^2}{\partial u_{1,s}\partial u_{2,s}}\Phi_{\rho_s}\left(\Phi^{-1}(u_{1,s}), \Phi^{-1}(u_{2,s})\right)$$

$$= \frac{1}{\sqrt{1-\rho_s^2}}\exp\left[\frac{\Phi^{-1}(u_{1,s})\Phi^{-1}(u_{2,s})\rho_s}{1-\rho_s^2} - \frac{\rho_s^2\left(\left(\Phi^{-1}(u_{1,s})\right)^2 + \left(\Phi^{-1}(u_{2,s})\right)^2\right)}{2(1-\rho_s^2)}\right]$$

$$c_s^{\mathrm{Gu}}(u_{1,s}, u_{2,s}) \tag{C2}$$

$$= \frac{\partial^2}{\partial u_{1,s}\partial u_{2,s}}e^{-\left((-\ln u_{1,s})^{\varsigma_s}+(-\ln u_{2,s})^{\varsigma_s}\right)^{\frac{1}{\varsigma_s}}} = \frac{u_{1,s}^{-1}u_{2,s}^{-1}}{\exp\left[\zeta^{\frac{1}{\varsigma_s}}\right]}\frac{1 + (\varsigma_s - 1)\,\zeta^{-\frac{1}{\varsigma_s}}}{(\ln u_{1,s}\ln u_{2,s})^{1-\varsigma_s}\,\zeta^{2-\frac{2}{\varsigma_s}}}$$

$$c_s^{\mathrm{Cl}}(u_{1,s}, u_{1,s}) \tag{C3}$$

$$= \frac{\partial^2}{\partial u_{1,s}\partial u_{2,s}}(u_{1,s}^{-\tau_s} + u_{2,s}^{-\tau_s} - 1)^{-\frac{1}{\tau_s}} = \frac{(1+\tau_s)(u_{1,s}u_{2,s})^{-1-\tau_s}}{(u_{1,s}^{-\tau_s} + u_{2,s}^{-\tau_s} - 1)^{2+\frac{1}{\tau_s}}}$$

$$c_s^{\mathrm{Pro}}(u_{1,s}, u_{1,s}) = \frac{\partial^2}{\partial u_{1,s}\partial u_{2,s}}u_{1,s}u_{2,s} = 1 \tag{C4}$$

where $\zeta = \left((-\ln u_{1,s})^{\varsigma_s} + (-\ln u_{2,s})^{\varsigma_s}\right)^{\frac{1}{\varsigma_s}}, s \in \{0,1\}$. The copula densities are necessitated for estimating dependence parameters as given in (5.12)-(5.14). With the empirical cdfs as in (5.11), the MLE of the copula parameters are then computed as

$$\hat{\rho}_s = \arg\max_{\rho_s}\sum_{n=1}^{N}\frac{2\,\Phi^{-1}(\hat{P}_{1,s}(x_{1,n}))\Phi^{-1}(\hat{P}_{2,s}(x_{2,n}))}{\rho_s^{-1}(1-\rho_s^2)} \tag{C5}$$

$$- \frac{\left(\Phi^{-1}(\hat{P}_{1,s}(x_{1,n}))\right)^2 + \left(\Phi^{-1}(\hat{P}_{2,s}(x_{2,n}))\right)^2}{1-\rho_s^2} - \ln\left(1-\rho_s^2\right)$$

$$\hat{\varsigma}_s = \arg\min_{\varsigma_s}\sum_{n=1}^{N}(1-\varsigma_s)\ln\left(\ln\hat{P}_{1,s}(x_{1,n})\ln\hat{P}_{2,s}(x_{2,n})\right) \tag{C6}$$

$$- \ln\left(1 + (\varsigma_s - 1)\,\hat{\zeta}^{-\frac{1}{\varsigma_s}}\right) + \ln\hat{\zeta}\left(2 - \frac{2}{\varsigma_s}\right) + \hat{\zeta}^{\frac{1}{\varsigma_s}} + \ln\left(\hat{P}_{1,s}(x_{1,n})\hat{P}_{2,s}(x_{2,n})\right)$$

$$\hat{\tau}_s = \arg\min_{\tau_s}\sum_{n=1}^{N}(1+\tau_s)\ln\left(\hat{P}_{1,s}(x_{1,n})\hat{P}_{2,s}(x_{2,n})\right) \tag{C7}$$

$$+ \ln\left(\left(\hat{P}_{1,s}(x_{1,n})\right)^{-\tau_s} + \left(\hat{P}_{2,s}(x_{2,n})\right)^{-\tau_s} - 1\right)^{2+\frac{1}{\tau_s}} - \ln\left(1+\tau_s\right)$$

where $\hat{\zeta} = \left((-\ln\hat{P}_{1,s}(x_{1,n}))^{\varsigma_s} + (-\ln\hat{P}_{2,s}(x_{2,n}))^{\varsigma_s}\right)^{\frac{1}{\varsigma_s}}$.

D   Detection Performance Limit

This appendix provides additional evaluation of the robust detector performance over different amounts of uncertainty in the distributions, which is realized by shifting the lower density bound $\underline{p}_s$, precisely by changing the constant value $\delta \in \{0.8, 0.7, 0.6, 0.59, 0.58, 0.57, 0.56, 0.55, 0.54, 0.53\}$, where $\underline{p}_s = \delta p_s$ and $s = \{0, 1\}$. Gaussian copula function is used to model statistical dependence between FL-GPR images from two consecutive viewpoints. Here, two different fusion rules are considered, namely, the majority voting ($\beta = 0.5$) and the hard fusion ($\beta = 1$) as given in (5.17). To provide a fair comparison, dependence structure is also incorporated into the parametric LRT, i.e., the LFDs in the test statistic in (5.8) are replaced by the parametric densities (p_0, p_1). With the parametric densities, the correlation coefficients were estimated as 0.94 and 0.99 for copula densities under hypotheses H_0 and H_1, respectively, which uses the estimation method of IFM [JX96]. The detection performance is evaluated based on the the number of MD and the number of FAs which are provided in Tables 1, 2 and 3 for the parametric test, robust test and robust test with lower nominal false alarm rate α, respectively. An overall evaluation of the detectors is provided below:

- As seen in Table 1, incorporating statistical dependence, $c_s \neq 1$, between tomographic images into the basic parametric test ($c_s = 1$) can evidently reduce the number of false alarms for both nominal false alarm rates and the fusion procedures β.

- As shown in Table 2, the copula-based robust test ($c_s \neq 1$) can detect all nine targets for δ from 0.56 to 0.8 with voting procedure ($\beta = 0.5$), while the robust test with no statistical dependent assumption ($c_s = 1$) always has one missed detection (target 9).

- For a lower pre-set α, the copula-based robust test with a 20% contamination of the distribution ($\delta = 0.8$) and the voting procedure is preferable. It only results in three false alarms and one missed detection as shown in Table 3. Moreover, under both dependent and independence assumptions, the voting procedure is preferred for the robust test with $\delta \geq 0.55$.

- The performance of the robust detectors depends on the amount of uncertainty in the distribution, characterized by δ. The δ value has to be chosen carefully before it achieves a breakdown point. For example, the copula-based robust test breaks down at $\delta = 0.54$, resulting in eight missed detections.

TABLE 1: Number of MDs and number of FAs resulted from *parametric* test.

| | $\alpha = 0.05$ | | | | $\alpha = 0.01$ | | | |
| | $\beta = 0.5$ | | $\beta = 1$ | | $\beta = 0.5$ | | $\beta = 1$ | |
	MD	FA	MD	FA	MD	FA	MD	FA
$c_s(\cdot) = 1$	0	48	0	30	4	1	5	1
$c_s(\cdot) \neq 1$	0	25	1	18	4	0	5	0

TABLE 2: Number of MDs and number of FAs resulted from *robust* test for $\alpha = 0.05$.

| | $c_s(\cdot) = 1$ | | | | $c_s(\cdot) \neq 1$ | | | |
| δ | $\beta = 0.5$ | | $\beta = 1$ | | $\beta = 0.5$ | | $\beta = 1$ | |
	MD	FA	MD	FA	MD	FA	MD	FA
0.8	1	20	1	12	0	44	0	34
0.7	1	15	1	10	0	43	0	25
0.6	1	13	1	9	0	38	0	19
0.59	1	11	1	8	0	36	0	18
0.58	1	11	1	7	0	30	1	16
0.57	1	11	1	7	0	27	1	12
0.56	1	11	1	7	0	20	1	9
0.55	1	11	1	7	1	12	1	6
0.54	1	11	1	5	2	3	7	0
0.53	9	0	9	0	8	0	8	0

TABLE 3: Number of MDs and number of FAs resulted from *robust* test for $\alpha = 0.01$.

| | $c_s(\cdot) = 1$ | | | | $c_s(\cdot) \neq 1$ | | | |
| δ | $\beta = 0.5$ | | $\beta = 1$ | | $\beta = 0.5$ | | $\beta = 1$ | |
	MD	FA	MD	FA	MD	FA	MD	FA
0.8	2	3	3	1	1	3	3	1
0.7	1	2	5	0	1	5	2	1
0.6	1	3	6	0	1	5	2	2
0.59	1	4	5	0	1	5	2	2
0.58	1	4	3	0	1	5	2	2
0.57	1	4	3	0	1	5	2	2
0.56	1	4	5	0	1	5	2	2
0.55	1	4	5	0	1	7	2	2
0.54	1	5	5	0	8	0	8	0
0.53	9	0	9	0	8	0	8	0

www.ingramcontent.com/pod-product-compliance
Lightning Source LLC
LaVergne TN
LVHW091622170726
843492LV00007B/2559